SpringerBriefs in Food, Health, and Nutrition

Springer Briefs in Food, Health, and Nutrition present concise summaries of cutting edge research and practical applications across a wide range of topics related to the field of food science, including its impact and relationship to health and nutrition. Subjects include: Food Chemistry, including analytical methods; ingredient functionality; physic-chemical aspects; thermodynamics Food Microbiology, including food safety; fermentation; foodborne pathogens; detection methods Food Process Engineering, including unit operations; mass transfer; heating, chilling and freezing; thermal and non-thermal processing, new technologies Food Physics, including material science; rheology, chewing/mastication Food Policy And applications to: Sensory Science Packaging Food Qualtiy Product Development We are especially interested in how these areas impact or are related to health and nutrition. Featuring compact volumes of 50 to 125 pages, the series covers a range of content from professional to academic. Typical topics might include:

- A timely report of state-of-the art analytical techniques
- A bridge between new research results, as published in journal articles, and a contextual literature review
- A snapshot of a hot or emerging topic
- An in-depth case study
- A presentation of core concepts that students must understand in order to make independent contributions

Shaba Noore • Shivani Pathania • Pablo Fuciños
Colm P. O'Donnell • Brijesh K. Tiwari

Nanocarriers for Controlled Release and Target Delivery of Bioactive Compounds

 Springer

Shaba Noore
School of Biosystems and Food Engineering
University College Dublin
Dublin, Ireland

Pablo Fuciños
Food Processing and Nutrition
International Iberian Nanotechnology
Laboratory
Braga, Portugal

Brijesh K. Tiwari
Department of Food Chemistry
and Technology
The Irish Agriculture and Food
Development Authority
Dublin, Ireland

Shivani Pathania
Food Industry Development Department
Teagasc - The Irish Agriculture and Food
Development Authority
Dublin, Ireland

Colm P. O'Donnell
School of Biosystems and Food Engineering
University College Dublin
Dublin, Ireland

ISSN 2197-571X   ISSN 2197-5728   (electronic)
SpringerBriefs in Food, Health, and Nutrition
ISBN 978-3-031-57487-0   ISBN 978-3-031-57488-7   (eBook)
https://doi.org/10.1007/978-3-031-57488-7

This Springer imprint is published by the registered company Springer Nature Switzerland AG
The registered company address is: Gewerbestrasse 11, 6330 Cham, Switzerland

Paper in this product is recyclable.

Preface

The functional food industry has long endeavored to discover effective means of safeguarding bioactive compounds from environmental degradation. The growing demand for high-quality bioactive compounds poses significant challenges for research, necessitating the advancement and modification of existing procedures, potentially including the replacement of current techniques applied in bioactive processing. Nanoencapsulation, a technique that involves encapsulating bioactive compounds using a suitable carrier, emerges as a solution to protect these compounds. It effectively preserves bioactives while inducing minimal changes in their color, texture, appearance, and nutritional value. Nanoencapsulation stands out as one of the most prominent recent innovations in processing, preservation, and delivery of bioactive compounds.

The present book comprises major sections on different types of nanocarriers used for encapsulating bioactive compounds. While unable to cover the entire field, the writing of this book was driven by the desire to compile and review the latest literature. The approach taken addresses not only the most recent applications but also novel and standard applications that have emerged in response to the challenges faced by the bioactives industry. This book also aims to provide an in-depth view of the controlled release and target delivery of nanoencapsulated bioactive compounds. This book provides valuable sources of information in the area of nanotechnology.

Dublin, Ireland   Shaba Noore
Dublin, Ireland   Shivani Pathania
Braga, Portugal   Pablo Fuciños
Dublin, Ireland   Colm P. O'Donnell
Dublin, Ireland   Brijesh K. Tiwari

Contents

Chapter 1
Introduction

Over the past few years, the demand for bioactive compounds offering health benefits has increased all around the globe. Bioactive compounds comprises numerous compounds such as polyphenols, alkaloids, carotenoids, terpenoids, glucosinolates, steroids, saponins, peptides, flavonoids, lignans, phytosterols, tannins, curcuminoids, quinones, phenolic acids and fatty acids, to name a few. These compounds are naturally found in a variety of sources including plants, fruits, vegetables, herbs, microorganicms, marine organisms, animals and insects (Benshitrit et al. 2012; McClements et al. 2009; Sagalowicz and Leser 2010). Bioacttive compounds are typically categorized as secondary metabolites, known for their potential in managing lifestyle disorders such as Alzheimer's disease, arthritis, cholesterol imbalances, asthma, cancer, and diabetes mellitus (DM). In nature, bioactive compounds are not found in their free form, necessitating the use of various techniques for their extraction and isolation to facilitate their targeted application. While conventional methods like soxhlet extraction, maceration, and hydro-distillation are employed commercially, there is a growing demand for sustainable, chemical-free alternatives due to the excessive use of solvents and longer processing times. Therefore, novel extraction techniques, including ultrasound-assisted, enzyme-assisted, microwave-assisted, pulsed electric field-assisted, supercritical fluid, and pressurized liquid extraction processes, are gaining attention. Moreover, compounds are further purified using methods such as thin-layer chromatography (TLC), column chromatography (CC), low-pressure liquid chromatography (LPLC), medium-pressure liquid chromatography (MPLC), and high-performance liquid chromatography (HPLC). Given the inherent instability of these compounds, it becomes imperative to protect them from environmental factors like temperature, light, pH fluctuations, and moisture changes to protect there bioactivity. Additionally, for these compounds to exert their desired biological effects within the human body, they must be stable and capable of surviving the harsh conditions of the gastrointestinal tract (GIT) and

S. Noore et al., *Nanocarriers for Controlled Release and Target Delivery of Bioactive Compounds*, SpringerBriefs in Food, Health, and Nutrition, https://doi.org/10.1007/978-3-031-57488-7_1

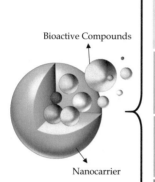

Bioactive Compounds

Food industry:
- Nanoemulsions and nanocarriers improve flavour and nutrient delivery.
- Nanosensors enhance food safety by detecting contaminants.
- Nanostructures extend shelf life through improved packaging materials.
- Nanoparticles enable targeted nutrient release in functional foods

Nutraceuticals:
- Nanoscale delivery systems enhance nutrient bioavailability.
- Nanocarriers enable controlled release of supplements in the body.
- Nanoencapsulation protects sensitive bioactives from degradation.
- Nanotechnology supports the development of personalized nutrition.

Pharmaceuticals:
- Nanoparticles improve drug solubility and bioavailability.
- Nanoencapsulation enables targeted drug delivery for precision medicine.
- Nanosensors monitor drug release and patient responses.
- Nanotechnology enhances drug formulation and stability.

Biotechnology and Medical Devices:
- Nanoscale materials are used in medical diagnostics.
- Nanoparticles assist in drug discovery and development.
- Nanotechnology supports tissue engineering and regenerative medicine.
- Nano-based medical devices improve diagnosis and treatment options.

Nanocarrier

Fig. 1.1 Application of nanoencapsualtion technology in food, nutracueticals, pharmacueticals and biotechnology field

stomach, necessitating the implementation of protective strategies to preserve their stability and bioavailability, allowing for controlled release and site-specific target delivery.

Nanoencapsulation is a protective strategy that has been effectively employed to create a protective shield around bioactive compounds. The significant role of nano-encapsulation in the food, nutraceuticals, pharmaceuticals, and biotechnology industries is illustrated in Fig. 1.1.

Nanotechnology for entrapping bioactive compounds utilizes appropriate nano-carriers resistant to enzymatic degradation from the mouth to the stomach and the gastrointestinal tract (GIT), thereby maximizing the proportion of protected active compounds that the intestine can effectively absorb. The primary nanocarriers employed in food systems include chitosan, zein, and alginate. (He and Hwang 2016; Maestrelli et al. 2006; Rashidi and Khosravi-Darani 2011). In the past decade, nanoencapsulation has evolved to such an extent that numerous techniques now exist for encapsulating various unstable compounds under typical environmental conditions. A variety of nanodelivery medium, including association colloids, bio-polymeric nanoparticles, nanoemulsions, nanofibers/nanotubes, and nanolaminates, are utilized to achieve the desired protective effect (Rashidi and Khosravi-Darani 2011).

The choice of a particular encapsulation technique depends on several factors, including the nature of the core and wall material, molecular size, thickness, solubility, permeability, and rate of delivery. These techniques can be categorized into three main genres: chemical methods (such as emulsion and interfacial polymerization), physical-chemical methods (including emulsification and coacervation), and physical-mechanical methods (like spray drying, spray cooling, spray congealing, prilling, freeze-drying, electrodynamic methods, and extrusion) (Perinelli et al. 2020). Reports indicate that approximately 80–90% of flavor encapsulation is achieved through spray drying, followed by 5–10% using spray chilling, 2–3% with melt extrusion, and around 2% by melt injection (Gupta et al. 2016b). Additionally, Coelho et al. (2021) highlights electro-spinning encapsulation as a heat-free technique for encapsulating fragrances, which shows promising results for heat-sensitive compounds (see Table 1.1). Electro-spinning serves as an ideal alternative to spray drying, yielding dry capsules without exposure to heat. However, in cases where obtaining a dry powder is not the goal, other techniques such as coacervation, emulsion, freeze-drying, spray chilling, etc., are utilized to achieve nanoencapsulation at low temperatures (Coelho et al. 2021) .

This review offers a comprehensive summary of various categories of nanocarriers used in the nanoencapsulation of bioactive compounds for potential applications in nutraceuticals, pharmaceuticals, and food formulations, drawing upon scientific literature published in the last 18 years. Notably, there has been a significant increase in the number of articles published on nanoencapsulation from 2017 to 2022, with the highest number of manuscripts published in 2022, as illustrated in Fig. 1.2. The primary objective of this review is to consolidate information on four categories of nanocarriers employed for encapsulating bioactive compounds. Additionally, it provides a brief overview of the bioactive compounds used in encapsulation and their applications in food systems. The review also outlines various opportunities and future challenges in this field.

1.1 Nanocarriers

The food industry widely employs nanoencapsulation techniques to safeguard bioactive compounds from environmental factors that can cause physical and chemical degradation. These techniques use specialized wall matrices known as nanocarriers. Nanocarriers have the dual function of preserving the natural flavor of bioactive compounds and masking the undesirable tastes and odors of certain compounds (Kiss 2020). Several matrices such as lipis, protein, carbohydrates and synthetic polymers are used as a nanocarried. Lipid based nanocarrier such as liposomes are among the most effective nanocarriers used for delivering bioactive compounds, as their lipid bilayer shell has been proven to protect the encapsulated material from degradation. Casein micelles have also been successful in encapsulating active components in dairy products, including minerals like phosphate and calcium (Shimoni 2009). Typically, nanoencapsulated materials have particle sizes ranging from 1 nm to less than 1000 nm, as shown in Fig. 1.3. Nanoencapsulation enables bioactive compounds to retain their functionality for extended periods and release from the encapsulant at significantly slower rates compared to nonencapsulated compounds.

Table 1.1 Comparison among various treatment techniques for nanoencapsulation

Strategies	Nanoparticles Size	Advantages	Disadvantages	Encapsulation Efficiency (%)	References
Ionic-gelation	40-100 nm	Less exposure to high temperature and vigorous stirring Minor solvent utilization Improved controlled release	–	91–95	Chaudhari et al. (2021), Naranjo-Durán et al. (2021)
High-pressure homogenization	~250 nm	Enhanced thermal stability Maintained crystallinity	Extensive use of energy	86	Md et al. (2019)
Ultrasonication	97–100 nm	Thermal stability of encapsulated nano- particles	Time-consuming Exposure of metal probe to the sample solution	82	da Silva Barbosa et al. (2021), Vasuki et al. (2014)
Electrospinning	287–997 nm	Enhancement in the bioactivity of the encapsulated material. Free from extreme temperature, pressure, and solvents.	Impossible to encapsulate water-soluble bioactive with the hydrophobic matrix	88–90	Li et al. (2021b), Rostamabad et al. (2020)
Electrospraying	<1 µm	Elimination of waste production Free from extreme temperature, pressure, and solvents. Low moisture powders	Low throughput	94	Đorđević and Đurović-Pejčev (2015), Wen et al. (2017)
Spray drying	<1 µm	Low cost of production Low moisture powders Long shelf life	Possibility of nozzle blockage during spray drying of high viscous solutions	76–95	Arpagaus et al. (2017)
Freeze drying	50–114 nm	Prevent bioactive compounds from degradation	Extensive use of energy Difficulties in scaling up	65–81	Ezhilarasi et al. (2013), Vahidmoghadam et al. (2019)

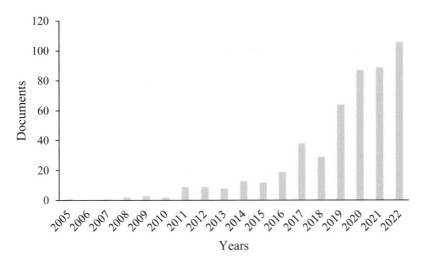

Fig. 1.2 Number of publications in last 18 years on nanoencapsulation of bioactive compounds

Moreover, various packaging industries, including those involved in meat and fruit packaging, utilize nanoencapsulation to extend the shelf life of their products by incorporating active nanoparticles into packaging materials (Liu et al. 2021).

Various matrices, including lipids, peptides, and carbohydrates, serve as nano-carriers for encapsulating bioactive compounds. These nanocarriers have been cat-egorized into two classes based on their application: food grade and non-food grade.

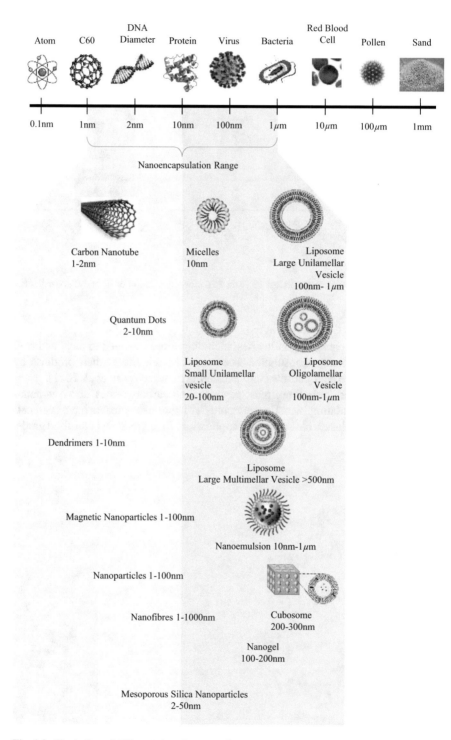

Fig. 1.3 Illustration of different size of nanocarriers

Chapter 2
Lipid-Based Nanocarriers

Lipid-based nanocarriers (LNCs) are designed from lipid molecules to encapsulate hydrophobic compounds such as polyphenols, lipophilic vitamins, aromas, and fatty acids. The previous generation of LNCs includes nanoemulsions, nanoliposomes, solid lipid nanoparticles (SLNPs), and nanostructured lipid carriers (NLCs). In recent years, there has been significant progress in LNC engineering, leading to the development of a new generation of LNCs, known as smart-lipid nano-carriers (SLNCs). These nanocarriers are primarily composed of lipids, which are naturally occurring or synthetic molecules that have hydrophobic and hydrophilic regions.

2.1 Nanoemulsions

Emulsions, in general, involve the combination of two immiscible liquids, where the internal or dispersed phase is distributed in drops within the external or continuous phase, forming a homogenous solution (Jafari et al. 2008). They are categorized into two primary types: single and double emulsions. Single emulsions can be further classified as oil-in-water (O/W) and water-in-oil (W/O) emulsions. Similarly, double emulsions come in two forms: oil-in-water-in-oil (O/W/O) and water-in-oil-in-water (W/O/W) (see Fig. 2.1). According to McClements (McClements 2004) emulsions with droplet sizes ranging from 100 to 1000 nm fall into the category of nanoemulsions.

The size and distribution of emulsion droplets play a vital role in determining emulsion quality, affecting factors such as stability, shelf life, color, texture, and appearance. Transparent emulsions typically have droplets ranging from 50 to 200 nm, while cloudiness may appear when droplets exceed 500 nm (McClements 2012). Nanoemulsions offer several advantages due to their nanoscale droplet size,

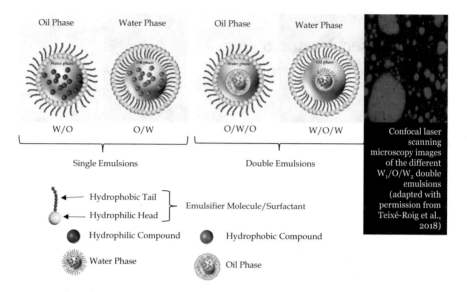

Fig. 2.1 Types of nanoemulsions. (Confocal laser scanning image adapted from (Teixé-Roig et al. 2018))

making them superior to emulsions with larger droplets (Esfanjani et al. 2015, 2017; Mohammadi et al. 2016a, b).

- Nanoemulsions experience reduced gravitational forces due to good Brownian motion, preventing sedimentation or creaming during storage.
- Nanodroplets inhibit flocculation (the conversion of nano-droplets into larger droplets due to attractive interactions), maintaining a dispersed network.
- Coalescence (the merging of two nano-droplets into larger ones) is constrained as nano-scale droplets are non-deformable, preventing surface modifications.
- Nanoemulsions require lower levels of surfactants compared to micro-emulsions, making them preferable for industrial use.
- Nanosized emulsions are considered safe from a toxicological perspective, encouraging industrial-scale production.
- Nanoemulsions are reported to offer enhanced nutrient bioavailability compared to micro-emulsions.

Research on nanoemulsions has expanded significantly in recent years, resulting in numerous publications on advanced concepts (Assadpour et al. 2016a, b; Mehrnia et al. 2016, 2017; Rao and McClements 2011). According to Scopus, there have been 114 research articles under the keyword search "nanoemulsions of bioactive compounds," with some of the latest findings presented in Table 2.1. For instance, bioactive compounds from pomegranate were encapsulated as a nanoemulsion using a 1:1 ratio of malt dextrin and whey protein isolate. The freeze-dried powder of this nanoemulsion was incorporated into soybean oil, resulting in improved physicochemical properties and enhanced antioxidant effects (Rashid et al. 2022).

Table 2.1 Nanoemulsions of bioactive compounds

Type of nanoemulsion	Nanocarriers	Emulsifiers	Bioactive compounds	Encapsulation strategies	Results	References
O/W	Gellan gum-based hydrogel	Span 60	Basil oil	Ultra sonication	The basil oil nano-emulsion hydrogels reported a 20% increase in ant biofilm activity than basil oil nano-emulsion.	Chinnaiyan et al. (2022)
O/W	Soybean protein isolate/ sodium alginate	–	Medium-chain triglyceride-oil	High-pressure homogenization	Addition of 0.02–0.06% (w/v) sodium alginate delayed the movement of oil droplets due to the greater ζ-potential among oil droplets. High-pressure homogenization at 100 MPa decreased the particle size of oil droplets and thus significantly improves the stability of nano-emulsion for application in food formulations.	Zhou et al. (2022)
W/O	Propylene glycol, glycerin, methylchloroisothiazolinone	Sorbitan monooleate/ Polysorbate 80	*Passiflora alata; Passiflora cincinnata; Passiflora setacea; Passiflora tenuifila*	Ultrasonication bath	After 48 h, the nano-emulsions showed higher cell proliferation in HaCaT keratinocytes treated cells. *Passiflora alata* nano emulsion and *Passiflora setacea* were superior to the control in concentrations above 12.5 μg/mL.	de Souza et al. (2022)

(continued)

Table 2.1 (continued)

Type of nanoemulsion	Nanocarriers	Emulsifiers	Bioactive compounds	Encapsulation strategies	Results	References
O/W/O	Tuna fish oil/medium-chain triglycerides	Purity gum ultra/Tween-80	β-Carotene/cinnamaldehyde and Vitamin E	Homogenization/ultrasonication	Purity gum ultra-based nano-emulsions presented more excellent β-carotene retention (42.3%) and Vitamin E retention (90.1%) over one-month storage at 40 °C than twee 80.	Ali et al. (2022)
W/O	Sesame oil or olive oil, or grapeseed oil	Tween 80	Vitamin D	Homogenization/high-pressure homogenization	Slow-release in encapsulated vitamin D (60%) compared to control (100%) in 24 h.	Jafarifar et al. (2022)
NSLCs	Precirol/acid oleic	Poloxamer 188 plus	Vitamin D	Homogenization/ultrasonication	Slow-release in encapsulated vitamin D (45%) compared to control (100%) in 24 h.	Jafarifar et al. (2022)
O/W	Corn starch	Tween 80	Amla essential oil	Ultrasonication	The fruits coated with nano-emulsions reflected significantly better quality fruit shelf life (15 days) compared with uncoated (7 days) Low degradation of bioactive compounds with high retention of antioxidant activity was also recorded.	Braich et al. (2022)

W/O	Glycerol and chitosan	Tween 80	*Zataria multiflora Boiss and Bunium persicum Boiss*	Ultrasonication	Chitosan-loaded nano-emulsion coating improved the oxidative stability of Turkey meat fillets. The samples' total volatile basic nitrogen level did not surpass the acceptability level (28–29 mg N/100 g).	Keykhosravy et al. (2022)
O/W	–	Tween 80	Curcumin/*Essential oil (Echium plantagineum)*	Homogenization/ microfluidization	Curcumin incorporation into nano-emulsions resulted in significantly higher antioxidant capacities (0.350 Torlox/g samples) than control samples (0.100 Torlox/g samples).	Inal et al. (2022)
W/O	–	Tween-80/ Span-80	Massoia lactone	Ultrasonication	Massoia lactone-loaded nano-emulsions reflected in stable anti-fungal activity against the pathogenic yeast *Metschnikowia bicuspidate* LLAO during storage (till 30 days). The diameters of the clear zones were 1.3 ± 0.1 cm and 1.3 ± 0.0 cm, respectively, in *Eriocheir Sinensis* for free Massoia lactone and Massoia lactone loaded in the nano-emulsions.	Yuan et al. (2022)

(continued)

Table 2.1 (continued)

Type of nanoemulsion	Nanocarriers	Emulsifiers	Bioactive compounds	Encapsulation strategies	Results	References
O/W	Zein	–	Essential oil (Mosla chinensis)	Homogenization	Incorporating essential oil into chitosan/zein-based film improved its antioxidant (from 30% to 60% of DPPH inhibition) and antibacterial properties (the zone of inhibition against S. aureus was improved from 12.40 mm to 26.62 mm).	Li et al. (2022)
W/O/W	Monoglyceride/olive oil	Pluronic F-68 + tween 80	Curcumin	High-speed homogenization/ultrasonication/freeze drying	An equal proportion mixture of PF-68+ Tw80 surfactants (1.25 or 2.5% each) improved the colloidal stability over time and provided desirable particle size values. For both Monoglyceride concentrations (10 and 35% w/w), the lowest oleogel/aqueous phase ratio (5/95) led to the formation of the optimum gelled-oil nano-particles that were stable for at least 303 days.	Palla et al. (2022)

| W/O/W | Maltodextrin and whey protein | Tween 80 | Pomegranate peel extract | Homogenization/ ultrasonication/ freeze drying | Controlled release of phenolic compounds provided oxidation protection in soybean and mustard oil. Oxidation protection by encapsulated pomegranate peel extract with complex maltodextrin/whey protein isolate was (20-meq oxygen/ kg) better than the un-encapsulated extract (70 meq oxygen/kg) in soybean oil. | Rashid et al. (2022) |

Similarly, an oleogel nanoemulsion was developed using monoglyceride (MG) to encapsulate curcumin, demonstrating higher encapsulation efficiency (91.95%) compared to curcumin encapsulated in a gelator (56.99%). Additionally, MG-encapsulated curcumin exhibited a slower release (Palla et al. 2022).

2.2   Nanoliposomes

Nanoliposomes serve as nanocarriers designed to encapsulate nutrients or pharmaceutical compounds, deriving their name from Greek words. 'Lipo' means fat, and 'soma' means structure/body. Consequently, nanoliposomes represent a structure of fat within which a compartment entraps bioactive compounds (Mozafari et al. 2008). Liposomes are nanoscale spherical vesicles made of phospholipids with an aqueous internal cavity. They come in various types depending on their formation process, size, and internal structure, with some having one or more concentric layers (Mozafari 2005). The formation of liposomes is driven by the interaction between aqueous and amphiphilic lipids. The hydrophilic head of the phospholipid faces the aqueous phase (internal phase), while the hydrophobic tail aligns with the bilayer. This dual-phase nature allows for the encapsulation of both hydrophilic and hydrophobic bioactive compounds (Fang et al. 2008).

Hydrophilic compounds are enclosed within the liposome's internal compartment, while hydrophobic compounds are trapped between the phospholipid bilayers. Liposomes offer several intriguing properties:

- Controlled release of encapsulated bioactive and medicinal compounds.
- Non-toxic and environmentally friendly.
- Enhanced bioavailability of nutrients.

The size of nanoliposomes typically falls within the range of 10 nm but may extend to micrometers (see Fig. 2.2) (Demirci et al. 2017).

Various techniques have been employed to prepare liposomes over the years, with abundant literature on optimizing their development conditions and techniques. Researchers have even developed advanced versions of liposomes with improved compound release mechanisms (Santana et al. 2013). Common techniques for liposome preparation include ultra-sonication, extrusion, micro-fluidization, and freeze-thawing (Maherani et al. 2012).

2.3   Cubosomes and Hexosomes

Cubosomes and hexosomes are nano-liquid crystal particles used as carriers for bioactive and pharmaceutical compounds, frequently employed for delivering bioactive compounds in functional foods. These structures exhibit thermodynamic stability and can encapsulate both hydrophobic and hydrophilic compounds. Cubosome

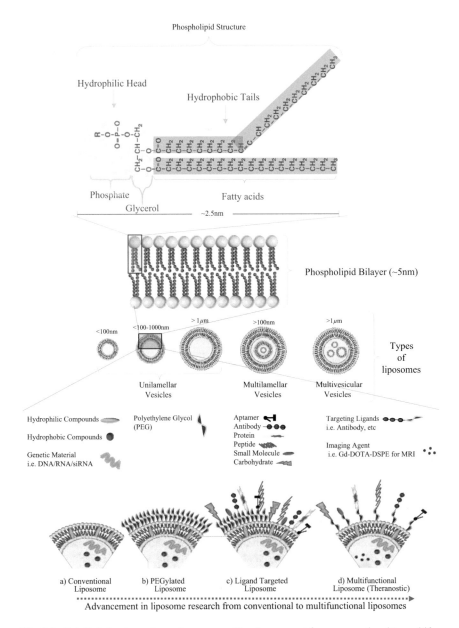

Fig. 2.2 Detailed structure of nanoliposomes with advancement from conventional to multifunctional liposomes

and hexosome crystals form when an amphiphilic lipid in the aqueous phase disperses, resulting in self-assembly due to its hydrophobic properties. This phase is known as the lyotropic crystal phase (LLC) and encompasses various categories,

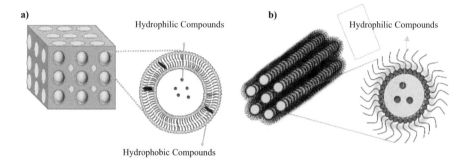

Fig. 2.3 Structure of (**a**) Cubosomes and (**b**) Hexosomes

including normal hexagonal, lamellar, micellar cubic, inverted double cubic, and inverted hexagonal phases, each distinguished by its structure (Tran et al. 2018).

The lamellar phase has an average interfacial curvature of zero, while the normal (O/W) phase has positive curvature and the inverse (W/O) phase has negative curvature. Diluting LLCs in an aqueous phase transforms them into spherical micelles, cylindrical micelles, and vesicles inverted crystals like cubosomes and hexosomes (Karami and Hamidi 2016; Yaghmur 2019).

Cubosomes, illustrated in Fig. 2.3a, are three-dimensional nanocrystals consisting of a bicontinuous lipid bilayer with a thickness of approximately 3.5 nm and two non-intersecting symmetrical water channels (Im3m, Pn3m, or Ia3d). Hexosomes (Fig. 2.3b), on the other hand, comprise filled water cylindrical micelles with tightly packed hexagonal cross-sections (Chen and Li 2021; Guo et al. 2010; Mertins et al. 2020). When used as delivery systems, hydrophilic compounds are located within the aqueous channels inside these crystals, while hydrophobic compounds are trapped in the lipid bilayer (Fan et al. 2021; Revathi and Dhanaraju 2014).

Compared to conventional lipid micelles, cubosomes and hexosomes offer higher compound loading capacities, larger surface areas, and improved chemical and mechanical attributes (Mo et al. 2017; Tenchov et al. 2021). Additionally, they can be easily activated by moderating temperature and pH, making them suitable for controlled and targeted delivery (Mertins et al. 2020; Zhai et al. 2019). However, these structures have limitations, such as difficulty in loading larger molecules due to small mesopores (4-6 nm), poor absorption in mucus, and low resistance in the gastrointestinal tract (GIT). These drawbacks can be addressed by encapsulating cubosomes and hexosomes with a surface coating, enhancing their applicability for targeted delivery (Chen and Li 2021; Tan et al. 2016a).

Numerous bioactive compounds, including vitamins, peptides, phenolics, and other nutraceuticals, have been successfully encapsulated using these strategies (Li et al. 2021a; Murgia et al. 2020). Commonly used amphiphilic lipids in preparing cubosomes and hexosomes include glycerol monooleate (GMO), monoelaidin, phytantriol (PHYT), phosphatidylethanolamine, oleoylethanolamine, phospholipids, glycolipids, and PEGylated phospholipids (Zabara and Mezzenga 2014). The basic structure of cubosomes and hexosomes is typically prepared using GMO and PHYT, known for their thermodynamic neutrality alongside excess water (Qiu and Caffrey 2000). While GMO is a naturally occurring lipid widely used in nutraceutical formulations, a study comparing GMO and PHYT cubosomes loaded with oridonin showed that PHYT cubosomes offer improved bioavailability and longer shelf life than GMO (Shi et al. 2017).

2.4   Tocosome

Tocosomes are colloidal or vesicular nanocarriers with significant constituents of phosphate group-bearing alpha-tocopherols (Mozafari et al. 2017). They can also accommodate various compounds like proteins, sterols, and polymers in their structure, similar to nano-liposomes. Alpha-tocopherol phosphate (TP) is naturally present in its phosphorylated form in food, human, and several animal tissues (Azzi 2006; Gianello et al. 2005). Recent literature indicates that TP is naturally found in green vegetables, cereals, fruits, nuts, seeds, and dairy products (Ogru et al. 2003). The general structure of TP comprises a phosphate group attached to one hydrophobic chain (phytyl tail) composed of three isoprene units. In the case of di-alpha-tocopherol (T2P), two phytyl chains are attached. However, TP cannot align like other phospholipids in a parallel bilayer structured position due to its bulky isoprene side-chain (Fig. 2.4). Therefore, the structure of the T2P molecule is conical, while the structure of TP is cylindrical (identical to the phosphatidylcholine molecule). Several clinical experiments have reported that TP and T2P molecules possess various health benefits, such as anti-inflammatory, cardioprotective attributes, and atherosclerotic-prevention properties (Libinaki et al. 2010; Munteanu et al. 2004).

Additionally, the anti-tumor effect of TP has also been cited (Saitoh et al. 2009). According to the literature, TP protects primary cortical neuronal cells from glutamate-induced cell toxicity in vitro and reduces lipid peroxidation in the plasma and liver of mice *in-vivo* (Nishio et al. 2011). In a recent study, TP and T2P were combined with different phospholipid and cholesterol molecules to encapsulate and target the release of the anticancer compound 5-fluorouracil (Mozafari et al. 2017). Due to its impressive health benefits, TP is widely used in food formulation and nutraceutical products.

2.5   Solid Lipid Nanoparticles

Solid lipid nanoparticles (SLNPs), as first-generation lipid particles, are fabricated by mixing solid lipid into the internal phase, while nano-lipid particles are developed by combining liquid and solid lipid (Fig. 2.5a). Chaudhari et al. (2021) encapsulated piperine and quercetin using Compritol as a solid lipid, squalene as the liquid lipid, and Span 80 and Tween 80 as emulsifiers and co-emulsifier. These

α- Tocopheryl Phosphate (TP)

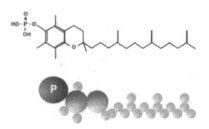

Molecular Formula: $C_{29}H_{51}O_5P$
Average mass: 510.69 Da
Monoisotopic mass: 510.347412 Da

(2R)-2,5,7,8-Tetramethyl-2-[(4R,8R)-4,8,12-trimethyltridecyl]-3,4-dihydro-2H-chromen-6-yl dihydrogen phosphate

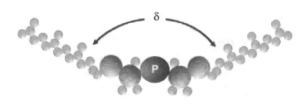

Di-α- Tocopheryl Phosphate (T$_2$P)

Molecular Formula: $C_{58}H_{99}O_6P$ Average mass: 923.54 Da

Fig. 2.4 Chemical structure of alpha tocopheryl phosphate (TP) and di-alpha tocopheryl phosphate (T2P). The angle of alignment of hydrocarbon chains (phytyl tails) of T2P molecule (δ) results in an "inverted truncated cone" molecule with a critical packing parameter (cpp) of greater than 1. (Adoted with permission from (Zarrabi et al. 2020))

encapsulated bioactive compounds exhibited slower release due to the slower erosion of the lipid wall matrix (12 h). Another study conducted by Abd-Elhakeem et al. demonstrated the improvement in bioavailability and oral targeted delivery of eplerenone using lipid-based nanoencapsulation. Eplerenone-loaded nano-lipid capsules showed enhanced permeability up to two times higher than the conventional aqueous drug in the rabbit intestine after 24 h.

2.6   Nanostructured Lipid Carriers

Nanostructured lipid carriers (NSLCs), Fig. 2.5b, are considered second-generation lipid nanostructures with advanced attributes compared to first-generation SLNPs. To prepare NSLCs, three main components are employed, including compatible (solid/liquid), biodegradable lipids, and surfactants/emulsifiers. Previously, liquid lipids were used to prepare SLNPs, resulting in several drawbacks, including leakage in the encapsulated structure and the release of bioactive compounds due to low

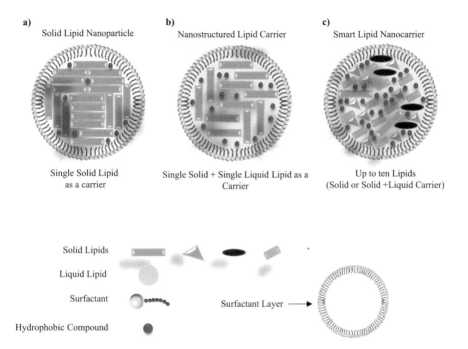

Fig. 2.5 Illustrated the basic structure of lipid based nanocarriers, (**a**) Solid lipid nanoparticle, (**b**) Nanostructured lipid carrier, (**c**) Smart lipid nanocarrier

crystalline structures caused by liquid lipids (Jain et al. 2017; López-García and Ganem-Rondero 2015). In recent years, the advanced structure of NSLCs has gained popularity among researchers, replacing SLNs, emulsions, liposomes, etc. (Jaiswal et al. 2016). These nano-scale NSLCs can encapsulate both hydrophilic and hydrophobic compounds, making them effective nanocarriers for drug delivery through various routes, including oral, ocular, topical, parenteral, and transdermal. They are also used for drug delivery to the brain and in chemotherapy. NSLCs have also found applications in cosmetics, food, and nutraceuticals (Jaiswal et al. 2016). Generally, there are three types of NSLCs depending on their morphological structure: crystalline, amorphous, and multi-model.

In the case of crystalline NSLCs, the matrix structure is disordered and comprises several void spaces to load more compounds in amorphous clusters. Solid lipids are mixed with liquid oils to achieve a perfect nano-crystal structure to overcome this drawback. However, due to the uneven chain of fatty acids and the combination of mono-, di-, and triacylglycerols, the ideal structure of NSLCs is not formed, but the mixing of different lipids enhances the loading capacity of compounds, albeit with a resulting minimum encapsulation efficiency (Iglič and Rappolt 2019; Selvamuthukumar and Velmurugan 2012).

In the case of multiple crystal structures of NSLCs, lipophilic drugs or bioactive compounds are less soluble in solid lipids and more soluble in liquid lipids. This results in multiple crystal-type NSLCs with enhanced liquid lipid content. A low concentration of oil moieties is effectively dispersed in the lipid matrix, while excess oil beyond its solubility in the solution results in phase separation, leading to small nano-compartments of oil encapsulated in a solid matrix. These multiple crystal structures are suitable for targeted drug delivery with reduced drug leakage (Ebrahimi et al. 2015; Iglič and Rappolt 2019).

In amorphous NSLCs, lipids are mixed using a controlled protocol to minimize drug leakage caused by crystallization during processing. Selected lipids, including hydroxyl stearate, isopropyl myristate, dibutyl adipate, and hydroxyl octacosanyl, possess characteristics that, during processing, form solid particles but are noncrystalline and thus create a homogeneous amorphous state(Ebrahimi et al. 2015; Iglič and Rappolt 2019; Selvamuthukumar and Velmurugan 2012).

2.7   Smart Lipid Nanocarriers

After using NSLCs, there was a drastic increase in the loading capacity of bioactive compounds inside the nano-lipid structures. For instance, the loading capacity of retinol enhanced from 1% to 5% in NSLCs compared to Smart lipid nanocarriers (SLNCs), Fig. 2.5c, respectively (Jenning et al. 2000). This increment in loading capacity also resulted in improved ordered structures. Results indicate a mechanism was certified that the loading capacity of the oil increases due to its better solubility in the oil phase, and a decrease in loading capacity takes place due to the reordering of the structures. Therefore, to avoid reordering in SLNCs, extremely chaotic lipid solutions are prepared using up to ten different blends of lipids (Müller and Keck 2015; Müller et al. 2014a, b; Py, et al. 2017). The stability of SLNCs was reported to be enhanced as SLNPs prepared using Dynasan 118 transformed into β-modification just after 1 month. In the case of SLNCs, after 1 year of storage, the solution remained in a primary α modification state under ambient conditions. Overall, SLNCs are considered an advanced version of NSLCs. The essential quality of the SLNCs matrix is that it comprises multiple solid lipids, or it can also contain one or additional oils. The chaotic mixture of lipids can also be prepared using NSLCs as it is already a mixture of several lipids such as Cutina LM. Cutina LM comprises fatty alcohols, waxes, and oils (Cetearyl Alcohol, Cetearyl glucoside, carnauba wax, candelilla wax, beeswax, and octyldodecanol). Hence, NSLCs are also called SLNCs when they are prepared with one solid lipid and a mixture of commercial lipids (Pyo et al. 2017).

Chapter 3
Protein-Based Nanocarriers

Protein-based nanoencapsulation possesses a higher compound/drug loading capacity than other nanostructures. It improves the absorption and bioavailability of the encapsulated compounds (Abaee et al. 2017; Chen et al. 2006). These nanostructures are prepared by the hydrophobic/hydrophilic interaction of bioactive compounds with the encapsulation matrices. Protein-based nanostructures are responsive to changes in the environment, such as pH change, temperature, enzymatic conditions, and ionic strength, making them suitable candidates for the targeted delivery of bioactive compounds to specified sites (Fang et al. 2014). Several types of proteins, such as whey, zein, and collagens, are used to form these nanocarriers. The release of these encapsulated compounds depends on their interaction with the encapsulation matrix; hydrophilic compounds are dispersed by diffusion, whereas hydrophobic compounds are released through enzymatic degradation of the protein matrix in the gastrointestinal tract (GIT). Additionally, these structures possess several limitations, such as disruption by the presence of protease enzymes in the GIT, making it a challenge to deliver bioactive compounds encapsulated with protein matrices (Bourbon et al. 2011; Donato-Capel et al. 2014). Nevertheless, there are different types of protein-based nanostructures, such as nanoparticles, nanohydrogels, nanotubes, hollow nanoparticles, nanofibrillar aggregates, electrospun nanofibers, and native state proteins as natural nanocarriers cited in the literature (Fig. 3.1) (Mohammadian et al. 2020).

3.1 Nanoparticles

Protein-based nanoparticles have gained substantial popularity as suitable nanocarriers for the encapsulation of bioactive compounds such as curcumin (Aniesrani Delfiya et al. 2016; Sadeghi et al. 2014; Teng et al. 2012), caffeine (Madadlou et al.

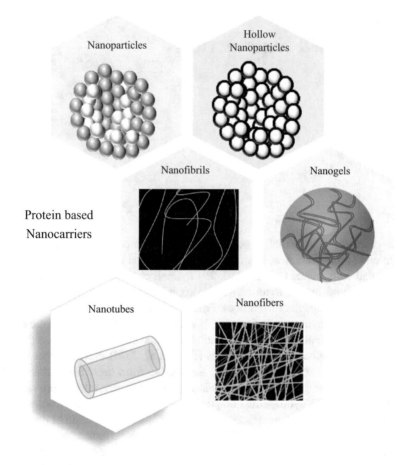

Fig. 3.1 Types of protein based nano-carriers

2014), vitamins (Abbasi et al. 2014; Wang et al. 2016a), resveratrol (Nowicka et al. 2019), carotenoids (Jain et al. 2018), etc.,. Several types of proteins are isolated from animal and plant sources, such as whey protein (Vickers 2017), BSA (Fang et al. 2011; Sadeghi et al. 2014; Yu et al. 2014), egg albumin (Aniesrani Delfiya et al. 2016), soy protein (Cheng et al. 2017; Teng et al. 2012; Wang et al. 2016a), zein (Cheng et al. 2017; Dong et al. 2016), etc. Additionally, techniques such as spray-drying, electrospraying, self-assembly, nano-precipitation, emulsification, coacervation, salting out, cross-linking, desolvation, etc., have been employed to fabricate nanoparticles (Jahanshahi and Babaei 2008; Tarhini et al. 2017). The selection of these strategies depends on the physicochemical properties of the encapsulation carrier and the molecular attributes of the encapsulation compound (Tarhini et al. 2017). The desolvation strategy has been used to fabricate protein-based nanoparticles due to its short preparation time and low degradation due to heating and shear rate (Fig. 3.2) (Vignoli et al. 2014).

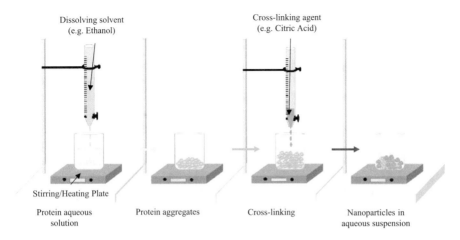

Dissolving solvent
(e.g. Ethanol)

Cross-linking agent
(e.g. Citric Acid)

Stirring/Heating Plate

Protein aqueous   Protein aggregates   Cross-linking   Nanoparticles in
solution   aqueous suspension

Fig. 3.2 Preparation of protein-based nanoparticles employing desolvation method

Briefly, dissolving agents like acetone or alcohol are added drop by drop to a protein solution with continuous stirring, leading to protein dehydration, thereby changing its conformation from stretched to coiled. These coils are further stabilized by cross-linking agents like citric acids, genipin, glutaraldehyde, etc. (Ekwall et al. 1990). Additionally, an opposite strategy involves adding a protein aqueous solution over an organic solvent, and in the case of zein, for example, the protein solubilized in ethanol water (approximately 90% ethanol) results in the formation of protein-based nanoparticles. Several studies have examined the release mechanism of these encapsulated bioactive compounds from the carrier matrix. Mechanisms include carrier degradation/erosion due to pH, temperature, or modification of enzymatic effects or release due to sonic or oscillatory magnetic field effects (Jahanshahi and Babaei 2008). According to reports, dissolving agents play an essential role in the shape formation of the nanoparticles during dissolution techniques. Small and spherical nanoparticles are prepared using ethanol, while acetone was used in rod-shaped particles in BSA protein as a nanocarrier (Sadeghi et al. 2014). Besides, the solubility of the encapsulation compound with the dissolving agent also affects the encapsulation process. For instance, the loading capacity of curcumin is low in nanoparticles prepared using ethanol compared to acetone, as curcumin is highly soluble in ethanol, preventing effective encapsulation.

3.2   Hollow Nanoparticles

Hollow nanoparticles are developed with interior void space, eliminating several disadvantages such as low-loading capacity, higher density, heat degradation, and limited surface area of solid nanoparticles (Wang et al. 2016b). Generally, these nanoparticles find applications in the nutraceutical delivery of targeted compounds.

These hollow structures are prepared in three steps: (1) the template is prepared as a core matrix, (2) the prepared template is coated layer-by-layer with a biopolymer coating, and (3) the removal of the inner template (Lou et al. 2008; Wang et al. 2016b). Several biopolymer matrices, such as lipids, proteins, and polysaccharides, are used to fabricate hollow nanoparticles. In the case of protein-based hollow nanoparticles, zein (Vickers 2017; Xu et al. 2011, 2013, 2015), collagens (Kraskiewicz et al. 2013), and casein proteins are employed. In one study, sodium carbonate crystals were used as a template for fabricating zein hollow nanoparticles (65 nm in size) to load an anti-diabetic drug. Further, the drug loading capacity was compared in both the nanoparticles, hollow, and solid zein nanoparticles. The results indicated that the hollow nanoparticles had a higher loading capacity than solid nanoparticles due to the large surface area and cavities developed during the release of the templates (Xu et al. 2011). Similarly, curcumin-loaded hollow nanoparticles were prepared using sodium carbonate as a core template and kafirin protein, extracted from sorghum flour, as shown in Fig. 3.3. The results of this study demonstrated efficient encapsulation of curcumin inside hollow kafirin nanoparticles. Additionally, hollow nanoparticles exhibited efficient controlled release of the encapsulated drug compared to solid nanoparticles (Li et al. 2019b).

Hollow nanoparticles can easily penetrate fibroblast cells, thereby delivering drugs such as peptides to the targeted cells. Cross-linked citric acid was employed to prepare hollow zein nanoparticles for the encapsulation of the anti-cancer drug known as 5-fluorouracil (Xu et al. 2015). The results indicated efficient loading and release of the drug at the targeted site and enhanced the enzymatic stability of the nanoparticles. Similarly, tannic acid was incorporated as a cross-linkage with zein

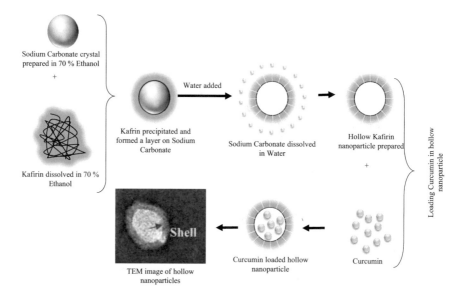

Fig. 3.3 Schematic representation for the preparation of protein based hollow nanoparticles

hollow nanoparticles for the oral delivery of curcumin, and the results indicated that cross-linking with tannic acid improved the enzymatic degradation of the zein matrix in the gastrointestinal tract (Vickers 2017).

3.3 Nanohydrogels

Nanohydrogels are three-dimensional nano-sized gel particles developed through the cross-linking of polymers. They possess a high capacity to retain water or biological fluids, exhibiting a substantial swelling ratio (Kwekkeboom et al. 2016; Zhang et al. 2016a). These nanogels are typically composed of natural or synthetic polymers, including edible proteins. Their larger internal space allows for increased loading capacity for bioactive compounds or drugs. Nanogels exhibit characteristics that enable them to penetrate living tissues, facilitating the targeted delivery of compounds (Abaee et al. 2017; Akiyama et al. 2007; Kwekkeboom et al. 2016; Zhang et al. 2016a).

The fabrication of nanogels relies on the self-assembly of proteins driven by physicochemical interactions, encompassing electrostatic, hydrophobic, hydrophilic, and amphiphilic associations (Zhang et al. 2016a). Several proteins such as soy protein (Chen et al. 2014; Feng et al. 2015; Jin et al. 2016), casein (Huppertz and de Kruif 2008), BSA (Wang et al. 2016c), gelatin (Akiyama et al. 2007; Wu and Wan 2008), lysozyme (Lin et al. 2015; Thakuria et al. 2013; Vickers 2017), and ovalbumin (Feng et al. 2016), etc. are employed in the preparation of nanogels. These nanogels find applications in encapsulating a wide range of compounds, including curcumin (Feng et al. 2015; Vickers 2017), caffeine (Bourbon et al. 2016a), riboflavin (Jin et al. 2016), doxorubicin (Wang et al. 2016c; Wu and Wan 2008).

Nanogels produced through self-assembly typically exhibit enhanced stability across a broad pH and salt range. In nanogels prepared using glycated proteins and the self-assembly technique, protein molecules are absorbed into the hydrophobic structure of bioactive molecules, while polysaccharide molecules on the exterior provide strong static repulsion, ensuring nanogel stability (Li et al. 2015). During the self-assembly process, proteins initially bind with polysaccharides, often high molecular weight and non-ionic polysaccharides like dextran. Subsequently, the resulting protein-polysaccharide complex is heated to its isoelectric point, yielding a stable nanogel through a heat-gelation process (Feng et al. 2015; Jin et al. 2016).

For instance, a stable and pH-sensitive soy protein-based nanogel was created using self-assembly in a heated (95 °C) denatured protein solution with a pH of 5.9 (Chen et al. 2014). The primary forces involved in forming soy-based nanogels are disulfide bonds and hydrophobic interactions. In another study, soy β-conglycinin-dextran nanogels were fabricated, featuring a hydrodynamic diameter of approximately 90 nm via self-assembly at the protein's isoelectric point (Feng et al. 2015). These nanogels demonstrated stability and suitability for the targeted delivery of encapsulated riboflavin. The electrostatic bonding between riboflavin's carboxyl groups and amino acids of the protein led to a 65.9% loading efficiency within the nanogels. Furthermore, *in-vitro* release of riboflavin within the gastrointestinal tract exhibited a gradual, controlled pattern, reducing potential side effects (Jin et al. 2016).

Egg proteins, such as ovalbumin and lysozyme, have also been utilized to create nanogels for encapsulating various bioactive compounds. Lysozyme, a globular protein, was combined with the biodegradable polysaccharide sodium carboxy-methyl cellulose to produce a nanogel (Thakuria et al. 2013). This nanogel was generated through heating and denaturation of the protein at 90 °C. It effectively encapsulated 5-fluorouracil, a widely used cancer treatment drug. The resulting nanogels had a particle size of approximately 241 nm and a loading capacity of 10.16% following 60 min of heating the nanogel solution. *In-vitro* studies demonstrated a slow and consistent release of the encapsulated drug within the gastrointestinal tract (GIT). Similar nanogels were employed to encapsulate the antitumor drug methotrexate (Li et al. 2015). Additionally, proteins sourced from animals and milk, such as lactoferrin, BSA, and glycomacropeptide, have been used to fabricate nanogels for encapsulating various antimicrobial and antioxidant compounds (Bourbon et al. 2015; Wang et al. 2016c). The key properties and hydrodynamic diameter of these protein-based nanogels depend on parameters like protein concentration in the nanogel (molar ratio), pH, heating temperature, and duration. Protein-based nanogels are considered an advanced encapsulation technique, often achieving encapsulation frequencies exceeding 90% (Bourbon et al. 2016a).

3.4   Nanofibrils

Several globular proteins are employed to fabricate nanofibrillar structures with dimensions ranging from 1–10 nm in thickness to 1–10 µm in length. These nanostructures are prepared through continuous heating of an acidic (pH 2.0) protein solution with low ionic strength. Generally, the formation of these structures is considered a self-assembling process (Mohammadian and Madadlou 2018; Serfert et al. 2014). Currently, nanofibrils have gained significant popularity in the food, medicine, and nanotechnology industries due to their remarkable attributes and functionalities (Farjami et al. 2016).

Various proteins, including those derived from milk, soy, rice, meat, eggs, and peas, are utilized in the fabrication of nano-fibrils. Two models, the monomeric and polypeptide models, are employed to create nano-fibrils through extensive heating at acidic pH (Mohammadian and Madadlou 2018). In the monomeric type, proteins are partially denatured, leading to the production of monomers. Nuclei are formed by the assembly of active monomers, followed by the growth and termination phase (Kroes-Nijboer et al. 2012). In this method, the production of low-active monomers during acid hydrolysis results in incomplete conversion of proteins into nanofibrils (Bolder et al. 2007). However, it has been reported that hydrolysis plays a vital role in fabricating protein-based fibrils. The peptides derived during hydrolysis are the building blocks of nanofibrils instead of complete/intact monomers (Akkermans et al. 2008). Therefore, the polypeptide method is preferred, as it produces fibril-forming peptides during hydrolysis (Mohammadian and Madadlou 2016). In both fabrication methods, nanofibers are developed in a sigmoidal pattern, beginning with the lag, elongation, growth phases, and concluding with the mutation phase (Kroes-Nijboer et al. 2012).

The structure of nanofibrils is influenced by several key parameters, including the initial protein concentration, processing time/temperature, ionic strength, and pH of the protein solution (Adamcik et al. 2010; Mohammadian and Madadlou 2016, 2018; Schleeger et al. 2013). Fabricated nano-fibrils are utilized for the encapsulation of bioactive substances, and notably, the hydrophobic surface properties of proteins increase during the nanofibril fabrication process. Consequently, these nanofibrils exhibit a heightened ability to bind with hydrophobic compounds.

According to reports, the solubility of curcumin was significantly enhanced by employing protein-based nanofibrils, achieving a 1200-fold increase compared to non-fibrillated whey proteins, which were limited to an 180-fold improvement (Mohammadian et al. 2019). Additionally, fibril-curcumin nano-complexes displayed enhanced antioxidant activity and slower compound release under simulated gastrointestinal conditions compared to non-fibrillated nano-structures.

3.5 Electrospun Nanofibers

Electrospun nanofibers are created using a high electrical force to draw from a polymer solution employing electrospinning equipment (Sullivan et al. 2014). This method of nanoencapsulation, used for fabricating nanofibers, is widely employed for heat-sensitive bioactive compounds because it doesn't involve a rise in temperature during the electrospinning process (Huang et al. 2013; Mendes et al. 2017; Tavassoli-Kafrani et al. 2018). The physicochemical attributes of these nanofibers depend on various electrospinning parameters, including flow rate, spinning distance, applied electrical power, and fluid-related factors like electrical conductivity, dielectric constant, surface tension, and viscosity (Mendes et al. 2017).

Proteins are an intriguing choice for fabricating nanofibers through electrospinning. However, during the spinning process, it's essential to mix a suitable carrier, such as poly(ethylene oxide), with the protein to enable efficient electrospinning and prevent the formation of protein capsules (Mendes et al. 2017). The addition of bipolymers also enhances the physicochemical properties of the nanofibers by reducing entanglement in the nanofiber chain (Zhong et al. 2018). Various food proteins, including soy, egg, zein, gelatin, and pea proteins, have been utilized to create electrospun nano-fibers for encapsulating bioactive compounds and drugs (Ghorani and Tucker 2015; Wongsasulak et al. 2010).

Multiple electrospinning modes are employed, such as blend, co-axial, and emulsion electrospinning, along with surface-modified electrospun fiber mats (Tavassoli-Kafrani et al. 2018). For example, whey protein conjugated with

spinnable biopolymers like poly(ethylene oxide) was utilized to fabricate electrospun nanofibers (Sullivan et al. 2014). These electrospun nanofibers, with diameters
ranging from 312 to 690 nm, were employed for encapsulating flavonoid compounds, specifically rhodamine B. The electrospun fibers created using a combination of protein and poly(ethylene oxide) showed uniform encapsulation of rhodamine
B compared to those made solely from poly(ethylene oxide). However, there was no
significant difference in the delivery/release of encapsulated rhodamine B; approximately 90% of the encapsulated compound was released after 10 minutes in the
aqueous phase.

In another study, Vitamin A and E were encapsulated in gelatin using electrospinning to fabricate nano-fibers suitable for wound dressing materials with antibacterial properties. This research revealed a delayed release profile of encapsulated
components for up to 60 h. Additionally, vitamins loaded in electrospun fibers
exhibited practical antibacterial effects against *Escherichia coli* (Li et al. 2016).
Similarly, in a separate study, essential oil extracted from orange fruit, employed as
a natural flavoring agent, was encapsulated in gelatin using an electrospinning
approach. Results indicated a two-fold enhancement in the stability of orange essential oil in electrospun nanofibers. Therefore, researchers concluded that electrospun
nanofibers based on gelatin are an excellent method for encapsulating bioactive
compounds in food and beverages (Tavassoli-Kafrani et al. 2018).

3.6   Nanotubes

Nanotubes represent a novel class of hollow tubular nanostructures crafted from
food proteins, serving as nanocarriers for encapsulating bioactive compounds.
These nanostructures find applications in various areas like gelation and viscosity
enhancement due to their distinctive attributes, including relative stiffness, nanoscale
cavities, high aspect ratio, and the capacity to form robust, transparent nanogels
(Graveland-Bikker et al. 2006; Tarhan and Harsa 2014).

Two primary methods for nanotube fabrication are widely utilized, involving
self-assembly of proteins/peptides to create hollow tube-like nanostructures and a
layer-by-layer electrostatic accumulation on template models (Katouzian and Jafari
2019; Sadeghi et al. 2013; Zhang et al. 2011). In the self-assembly method,
α-lactalbumin (whey protein) stands out as the sole food protein capable of forming
nanotubes after a specific degree of enzymatic hydrolysis (Geng et al. 2016). The
fabrication of nanotubes from α-lactalbumin involves three major steps: (1) protein
hydrolysis (partial) utilizing serine protease; (2) the formation of dimeric building
blocks at a high concentration in the presence of divalent cations (e.g., Ca2+),
resulting in the formation of bridges among the building blocks and the production
of a stable nucleus composed of five building units, and (3) elongation to enlarge the
size of tubular nano-structures (Graveland-Bikker et al. 2006). Several factors, such
as protein concentration, temperature, enzyme type/concentration, divalent cation

concentration, etc., significantly influence the formation of α-lactalbumin nano-tubes (Ramos et al. 2017).

In another study, chemical methods for protein hydrolysis were explored as alternatives to enzymes in the preparation of α-lactalbumin-based nanotubes Instead of enzymes, chemicals like surfactants, polar solvents, and acidic buffers were employed for protein hydrolysis. This chemical method was reported as a cost-effective and efficient approach without the need for high-temperature conditions. The nanotubes produced through this method exhibited diameters ranging from 3 to 8 nm (outer diameter) and found applications in food, medicine, pharmaceuticals, and nutraceuticals. The literature suggests that the growth rate of nanotubes has been studied. In one study, tubular nano-structures with a cylindrical diameter of 19.9 nm and a cavity diameter of 8.7 nm were created using α-lactalbumin at a concentration of 28 g/L, a pH of 7.5, and an enzyme-to-substrate molar ratio of 1:260. The calcium concentration was set at 3 mol/L of protein. The results indicated a growth rate (size elongation of nanotubes) of 10 nm/min. In another report, functional properties of α-lactalbumin nanotubes were investigated to assess their potential applications. The results demonstrated a significant ($p < 0.05$) increase in bioactivity, with an IC50 (protein concentration to inhibit 50% of ACE activity) of α-lactalbumin nanotubes at 281 ± 21 μg mL $-$ 1 and an antioxidant activity of 1.28 ± 0.01 μmol TE mg $-$ 1 protein (Fuciños et al. 2021). Additionally, a previous study by the same author on the stability and functionality of α-lactalbumin nano-tubes during freeze-drying revealed that at 8 °C and pH 7.5, > 50% of the caffeine remained encapsulated within α-lactalbumin nanotubes (Fuciños et al. 2017).

Furthermore, it was reported that due to high viscosity and gel-like properties, only 60% of the protein assembled into nanotubes (Graveland-Bikker et al. 2006). Circular dichroism spectroscopy was employed to confirm nanotube fabrication, revealing no significant changes in the protein's secondary structures during the process. Additionally, protein hydrolysates were found to form nanotubes with a molar mass of approximately 11 kDa (Graveland-Bikker et al. 2009). Another imaging technique known as Fourier transforms infrared spectroscopy (FTIR) unveiled that Ca^{2+} ions played a pivotal role in forming bridges among the negatively charged carboxyl groups attached to the Asp and Glu chains of peptides liberated during partial protein hydrolysis, leading to nanotube formation (Tarhan et al. 2014).

3.7 Natural Nanocarriers

In addition to various fabricated nano-structures, native proteins exhibit the inherent capability to encapsulate bioactives within their natural structure, forming nano-complexes. These nano-complexes play a crucial role in enhancing the solubility, adsorption, stability, and bioavailability of the encapsulated lipophilic compounds (Maldonado and Kokini 2018). Food proteins, in general, are naturally present in nano-sized forms, making them widely employed as nano-carriers for drug and bio-active compound encapsulation (Semo et al. 2007).

Bioactive compounds establish binding interactions with the lipophilic regions of native proteins, particularly through interactions with their aromatic rings, resulting in the formation of nano-complexes. Additionally, hydrogen bonds have been reported to contribute to the development of protein-bioactive compound nano-complexes (Liu et al. 2017; Tapal and Tiku 2012).

The process of nano-complexation is relatively straightforward: lipophilic bioactive compounds are initially dissolved in a suitable solvent (e.g., ethanol) and then mixed with the protein solution. These mixtures are stirred to induce the formation of nano-complexes. Subsequently, centrifugation is employed to separate non-encapsulated bioactive compounds, and the resulting nano-complexes can be further converted into powders through freeze-drying (Chen et al. 2015; Esmaili et al. 2011; Liu et al. 2018a).

During the formation of nano-complexes, it is crucial to use a solvent with a low concentration initially to avoid any impact on the physicochemical properties of the nano-carrier protein (Mohammadian et al. 2019). Various types of proteins isolated from both plants and animals, such as milk, pea, soy, and egg proteins, are utilized for the creation of nano-complexes (Jiang et al. 2019; Pujara et al. 2017; Visentini et al. 2017; Zhan et al. 2020) to encapsulate bioactive compounds such as curcumin (Chen et al. 2015), resveratrol (Liu et al. 2018a), vitamins (Jiang et al. 2019), gallic acid (Zhan et al. 2020), 1-Octacosanol (Li et al. 2019a), glabridin (Wei et al. 2018) and more. These bioactive compounds, when encapsulated in nano-complexes, exhibit significantly enhanced aqueous solubility, stability, improved bioavailability, and increased antioxidant activity. For instance, nano-complexation employing ovalbumin, β-casein, and soy protein for curcumin encapsulation resulted in a 370-fold, 2500-fold, and 812-fold increase in aqueous solubility, respectively (Esmaili et al. 2011; Liu et al. 2018b; Tapal and Tiku 2012). Consequently, nano-complexes are regarded as an efficient encapsulation approach for lipophilic bioactive compounds.

Casein, a major milk protein, is also extensively used as a natural nano-carrier for encapsulating bioactive compounds. In milk, casein exists in the form of micelles with an average colloidal size of 150 nm (Semo et al. 2007). Micellar nano-structures are typically formed through the accumulation of casein fractions via lipophilic interactions and calcium phosphate bridges (Ghayour et al. 2019). Caseins are considered natural nano-capsules, responsible for delivering essential nutrients like protein, phosphate, and calcium to infants (Semo et al. 2007), (Ghayour et al. 2019; Rehan et al. 2019; Sáiz-Abajo et al. 2013). Additionally, caseins can be fabricated in vitro to replicate the attributes of natural caseins (Sáiz-Abajo et al. 2013). In a specific study, vitamin D2 was encapsulated within fabricated casein micelles. Initially, vitamin D2 was mixed with a sodium caseinate solution and subsequently added to tri-potassium citrate, K2HPO4, and $CaCl_2$ under conditions of pH 6.7–7.0 and 37 °C. The resulting solution was continuously stirred, followed by centrifugation. The supernatant was separated, ultra-filtered, and subjected to high-speed homogenization. This study demonstrated an increased vitamin concentration within micelles, up to 5.5 times higher compared to serum levels. Furthermore, encapsulated vitamin D2 exhibited enhanced stability with minimal degradation when exposed to UV light (Semo et al. 2007).

3.8 Cage-Like Proteins Nanoencapsulation

The term "cage-like protein encapsulation" refers to a self-assembled encapsulation within a three-dimensional structure characterized by a hollow cavity. Examples of proteins exhibiting nano-cage structures include ferritin (Li et al. 2020), heat shock proteins (Kim et al. 1998), and DNA-binding protein (Suzuki et al. 2021). Typically, these structures are spherical and result from the self-assembly of multiple subunits into intricate super-molecular structures. The nano-protein cages consist of three distinct regions: the outer surface, the inner surface, and the interfaces between subunits (Lv et al. 2021).

The inner surface of these nano-cages resembles a sealed chamber used for encapsulating cargo compounds. This unique quality makes protein cages effective for bioactive encapsulation. The outer surface of these nano-cages is closely associated with cellular interactions, facilitating the targeted delivery of the encapsulated compounds (Montemiglio et al. 2019). Moreover, the outer surface can be genetically and chemically functionalized to enhance the stability of the nano-cage in the gastrointestinal environment and achieve targeted delivery of encapsulated compounds (Meng et al. 2019).

The interfaces among subunits play a crucial role in the formation of protein nano-cages. The arrangement of subunits and their interactions at the interfaces significantly influence the final structure of the protein cage (Zhang et al. 2016b). These nano-cages are organized symmetrically in various geometric shapes and consist of multiple pores that connect the inner cavity with the external environment. These pores facilitate the transfer of active compounds from the inner cavity for targeted delivery (Jiang et al. 2020). Natural protein cages are of several categories based on their shape, size, chemical stability, and structural morphology. Natural protein cages can be categorized based on their shape, size, chemical stability, and structural morphology.

3.9 Ferritin Nanocages

The Ferritin protein nano-cage is found widely in all forms of life except yeast and can be extracted from various plant and animal sources, including meat, soybeans, peas, and seafood. Ferritin nano-cages primarily serve as storage containers for iron ions, storing them in the form of iron oxyhydroxide. The structure of Ferritin consists of 24 subunits self-assembled into a dodecameric nano-cage with F432 symmetrical arrangement, featuring inner and outer diameters of approximately 8 and

12 nm, respectively (Lawson et al. 1991). Plant-based Ferritins, in particular, are considered iron supplements due to the iron content in their inner core (Gesinde et al. 2018). This inner iron core can be easily removed using appropriate reagents, resulting in hollow Ferritin without an inner iron core, known as apo-ferritin (Jutz et al. 2015).

Additionally, Ferritin exhibits C3/C4 symmetry axes, six hydrophobic channels, and eight hourglass-shaped hydrophilic channels with pore sizes ranging from 0.03 to 0.5 nm, all connected to the inner cavity of the nano-cage. There are also eight three-fold channels identified as pathways for iron entry into the nano-cage (Fig. 3.4). These nano-cage proteins can withstand high temperatures, up to 80 °C for 10 min. They are also capable of tolerating exposure to a wide range of pH levels, ranging from 3.0 to 10.0, and high concentrations of denaturants, including guanidine (Zhang et al. 2019b). Ferritin can be extracted from various plant and animal matrices, making it a suitable carrier for cargo delivery (Yang et al. 2015).

3.10 Dps Nanocage

The Dps nanocage, also known as a DNA-binding protein belonging to starved cells (DPs), is commonly found in microorganisms. Its primary function is to protect cells from oxidative stress and nutritional scarcity (Suzuki et al. 2021). The basic structure consists of 12 identical subunits arranged in a cage-like architecture with 23-fold symmetry (Ilari et al. 1999). Furthermore, these subunits are composed of four-helix structures, similar to ferritin. The resulting nano-cage has an inner diameter of 5 nm and an outer diameter of 9 nm, with thermal stability up to 70 °C and tolerance to a wide range of pH values (Minato et al. 2020). Therefore, this nano-cage can be utilized for the delivery of nanoparticles.

3.11 Heat Shock Protein

Proteins secreted by cells when exposed to stress induced by heat are termed heat shock proteins. Several heat shock proteins are widely found in living systems, including plants, animals, and microorganisms. These proteins act as chaperones responsible for protein folding and stabilization. They are further divided into different categories based on their size, including HPS 100, HSP 90, HSP 70, HSP 60, and small HSP (Dubrez et al. 2020). Generally, they are spherical and comprise 24 subunits of protein assembled in a hollow cage with octahedral symmetry, with outer and inner diameters of 12 nm and 6.5 nm, respectively. This structure is employed to encapsulate small molecules (Kim et al. 1998).

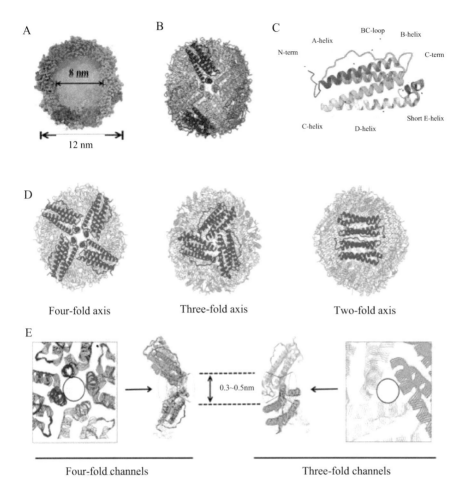

Fig. 3.4 (**a**) The inner and outer diameters of ferritin nano-cages. (**b**) The structure of ferritin assembled from 24 subunits in a F432 symmetry manner. (**c**) Ribbon representation of the structure of human H type ferritin subunit (PDB:2FHA). (**d**) Four-fold axis, three-fold axis, and two-fold axis of ferritin nano-cage. (**e**) The pore size of four-fold channels and three-fold channels is 0.3–0.5 nm. (Adapted from Chen et al. 2021)

3.12 Encapsulation Protein

Encapsulins are protein cages primarily found in the genomes of bacteria and archaea. These cellular protein cages play a crucial role in encapsulating cargo proteins that possess specific structures at their C-termini (Giessen and Silver 2017). High-resolution imaging studies of encapsulins from Thermotoga Maritima, *Pyrococcus furiosus,* and *Myxococcus Xanthus* have revealed that they assemble into icosahedral structures of various sizes. For example, Thermotoga Maritima encapsulins consist of 60 subunits and form T = 1 icosahedra with a diameter of

approximately 24 nm, while Pyrococcus furiosus and Myxococcus Xanthus encapsulins consist of 180 subunits and create T = 3 icosahedra with a diameter of roughly
32 nm (Akita et al. 2007; McHugh et al. 2014; Sutter et al. 2008).

The structural assembly of these subunits includes three primary openings or
pores (measuring 5–6 Å) arranged in two, three, and fivefold symmetries. These
openings are responsible for facilitating the exchange of molecules between the
inner cavity and the cytosol. Over the years, various pieces of evidence have
emerged regarding cargo proteins that can be enclosed within the inner compartment of encapsulins (Jones and Giessen 2021).

3.13   Pyruvate Dehydrogenase

Pyruvate dehydrogenase is a large enzyme complex that serves as a catalyst for
converting pyruvate into acetyl-CoA and reducing NAD+. It plays a crucial role in
both glycolysis and the tricarboxylic acid cycle (Milne et al. 2006). This multi-
complex enzyme consists of three main catalytic components: pyruvate decarboxyl-
ase (E1), dihydrolipoamide acetyltransferase (E2), and dihydrolipoamide
dehydrogenase (E3). The hollow-core structure is formed by the E2 protein, which
is primarily responsible for the chemical reactions (Mattevi et. al. 1992). In
Escherichia coli, the E2 protein is composed of 24 subunits, capable of forming a
self-assembled cage-like structure with a cubic core. In the case of Geobacillus
stearothermophilus, it is composed of 60-mer E2 proteins, forming an icosahedral
structure (Milne et al. 2006). Additionally, it has been reported that E2 proteins can
be produced independently of E1 and E3 proteins, maintaining similar stability and
native structure in vitro. Moreover, they are amenable to chemical and genetic modifications (Ren et al. 2012). Consequently, E2 proteins have gained popularity in
applications involving the encapsulation of specific compounds (Ren et al. 2012).

As mentioned earlier, the application of protein nano-cages, whether natural or
fabricated, varies based on their functionalities. However, a fundamental similarity
among all these structures is that they contain a cage-like inner cavity capable of
encapsulating compounds, isolating them from the outer environment. The inner
space of these nano-cages serves as a compartment for the storage and transporta-
tion of cargo molecules in vivo. Recently, protein-cage encapsulation has gained
significant prominence in nutraceutical and pharmaceutical sciences. Natural nano-
cages, such as E2 proteins and heat shock proteins, are utilized as nano-carriers for
delivering encapsulated compounds, including doxorubicin trapped in E2. The
porous openings in E3 provide a clear pathway for the compounds to penetrate
inside the nano-cage. Furthermore, the interaction between the compound and the
inner cavity confirms the entrapment of the compound within the nano-cage (Ren
et al. 2011).

In recent years, fabricated nano-cages have gained tremendous importance for
delivering bioactive compounds using protein folding and genetically engineered
methods and defines (Nasu et al. 2021). For instance, a 16-mer nano-cage was

fabricated to encapsulate food nutrients based on the natural ferritin protein nano-structure, featuring 8-mer concave discs and a nano-cage measuring 10 nm in length and 8 nm in width, respectively (Zhang et al. 2016b). Additionally, this nano-cage can be disassembled and reassembled based on the attributes of subunit interaction and pH stimuli from the outer environment. Generally, ferritin-type nano-cages disassemble at a pH value of around 4.0. A report illustrated that curcumin was encapsulated in a lenticular nano-cage with compound release triggered at a pH of 4.0. This type of cargo release also minimizes the damage to compounds in acidic pH conditions (Chen et al. 2016; Wang et al. 2019; Zhang et al. 2016b). Nevertheless, the stability of this protein-based nano-cage in the gastrointestinal tract (GIT) remains a challenge. Further research is needed to enhance its absorption, reduce cell cytotoxicity, improve bioavailability, and enable large-scale production at a lower cost.

.

Chapter 4
Carbohydrate Based Nanocarriers

After lipids and proteins, carbohydrates are the most abundant primary metabolites present in our environment. Mostly, carbohydrates exist in their polysaccharide form, with more than ten monosaccharide groups attached in glycosidic bonds. Cellulose and chitin are considered the most widely available homopolysaccharides, specifically in cell walls in plants and exoskeleton parts of arthropods such as insects, lobsters, crabs, and shellfish. Generally, heteropolysaccharides, including hyaluronate, chondroitin sulfate, and keratin sulfate found in cells, tissues, and organs, provide support, protection, and shape to bacteria and mammals (Glasser 2008).

Due to their physicochemical attributes, such as biodegradability, stability, safety, adhesiveness, hydrophilicity, and biocompatibility, polysaccharides are considered the best natural polymers among all other polymers for fabricating nano-carriers for the delivery of encapsulated compounds (Liu et al. 2008). These polysaccharides play a vital role in preparation as nano-carriers. They possess multiple functions, such as rheology control, surface modification, and emulsion stabilization, depending on the type of functional groups attached to the molecular structure of polysaccharides. Therefore, they can be tailored by chemically modifying the carbohydrate structure. Chemical modifications include carboxymethylation, grafting, etc. Carbohydrate-based encapsulation also depends on its polyelectrolytic charges, such as chitosan having positive charges, whereas alginate, pectin, hyaluronic acids, heparin, etc., comprise negative charges (Liu et al. 2008).

Specifically, chitosan and alginates have emerged as suitable carriers for delivering nutraceuticals due to their non-toxic, biodegradable, and biocompatible nature. They can protect bioactive ingredients from extreme conditions such as pH and temperature. Chitosan and alginate capsules have proven to be suitable carriers for essential oils, flavors, vitamins, antioxidants, and probiotics (Maleki et al. 2022). Additionally, carbohydrate-based nanoparticles are substrates for cell surface receptors, enabling site-specific delivery systems (Carrillo-Conde et al. 2011).

S. Noore et al., *Nanocarriers for Controlled Release and Target Delivery of Bioactive Compounds*, SpringerBriefs in Food, Health, and Nutrition, https://doi.org/10.1007/978-3-031-57488-7_4

These nanoparticles are fabricated employing various techniques such as emulsion droplet formation, ionic or covalent cross-linkage, self-assembly, nano-precipitation, or a combination of these strategies. Several polysaccharides tend to thicken upon dissolving in an aqueous phase, and due to their thickening ability, the gelation method is primarily preferred for the fabrication of polysaccharide-based nanoparticles. However, the preparation of nanoparticles depends on the molecular structure of the polysaccharides. In some cases, a lipophilic gelling agent is mixed into the aqueous phase before emulsification in the oil phase to ensure the formation of nanoparticles of the proper size (30–300 nm). Nanoparticles of chitosan (Yoksan et al. 2010), dextran (Rei, et al. 2007), and alginates (Hosseini et al. 2013) are fabricated by the above mentions method.

The release mechanism of these nanoparticles is based on the cross-linkage density within the molecular structure of these nanoparticles. This chemical reaction leads to a swelling network of the nanoparticle, resulting in the release of encapsulated compounds. Commonly used cross-linking agents include glutaraldehyde, dialdehydes, genipic, epichlorohydrin, etc. (Liu et al. 2008). Additionally, several carbohydrate-based nano-carriers depend on their physicochemical properties and encapsulation efficiency.

4.1 Polymeric Nanoparticles

Polymeric nanoparticles (NPs) are colloidal particles of submicron size with exquisite characteristics for encapsulating bioactive compounds, drugs, and genes (Mahapatro and Singh; Panyam and Labhasetwar 2003). Due to their low toxicity, high biodegradability, and biocompatibility, these nanoparticles are considered efficient vehicles for targeted delivery. Polymer-based nanoparticles can be easily altered or modified to create a multifunctional delivery system. For example, their shape, size, surface area, etc., can be controlled, along with their degradation or dispersion kinetics, to achieve efficient release of the encapsulated compound (Albertsson and Varma 2002). These nanoparticles are exceptionally stable and can effectively encapsulate both hydrophilic and lipophilic molecules with excellent encapsulation efficiency (Gelperina et al. 2005). These polymeric wall matrices not only protect the encapsulated compound from degradation but also enable penetration through extra and intracellular cell barriers due to their nano-sized particles, resulting in site-specific delivery of the encapsulated compounds, such as drug release near tumor cells after penetrating the endothelium (Prokop and Davidson 2008; Singh and Lillard Jr 2009).

Over the years, two types of polymeric nanoparticles, namely nanospheres and nanocapsules, have been employed as efficient nano-carriers for the entrapment of compounds. Nanospheres involve the homogeneous scattering of the encapsulated molecule in a polymeric matrix, while in the case of nanocapsules, the compound is entrapped inside the vesicular system of a polymeric wall material covering the core with the cargo compound (Singh and Lillard Jr 2009). In recent years, several techniques such as spray-drying, nano-precipitation, emulsions, salting out, etc., have been implemented to fabricate polymeric nanoparticles. The choice of these methods plays a vital role in determining the physicochemical properties of the fabricated nanoparticles (Görner et al. 1999; Lassalle and Ferreira 2007). Several polymers from various sources, such as chitosan, polylactic acid, poly-lactic-co-glycolic acid (PLGA), etc., have been used to fabricate polymeric nanoparticles (Krishnamachari et al. 2011). These nanoparticles efficiently preserve the immuno-genicity and antigenicity of the cargo proteins. PLGA is extensively used in vaccine formulation as antigens encapsulated in PLGA have shown improved cellular responses and antibody production compared to un-encapsulated antigens.

Additionally, if antigens cannot induce DC activation, modified nanoparticles comprising maturation signals on their surfaces can interact directly with ligand receptors. As a result, mannose receptors are overexpressed on the surfaces of mac-rophages and DCs. For instance, chitosan-based nanoparticles are incredibly effi-cient for encapsulating small interfering RNA (siRNA) genes. Due to the electrostatic interaction between the positive charge of chitosan and the negative charge on siRNA, the nanoparticle can be easily transported to its targeted site *in vivo* (Aslan et al. 2013). Furthermore, in Costa et al. (2021) encapsulated bioactive grape pom-ace extract using chitosan and alginate nanoparticles. The results indicated protec-tion of bioactive compounds from hydrolysis in the gastrointestinal tract with enhanced bioactivity.

4.2 Polymeric Micelles

Polymeric micelles (PMs) are spherically shaped nanoparticles with sizes ranging from 10 to 100 nm. They are prepared through the self-assembly of copolymers (amphiphilic blocks) in the aqueous phase. Generally, these nanomicelles are used for delivering hydrophobic compounds and drugs (Torchilin 2001, 2008). It has also been reported that these nanoparticles possess the ability to improve the bioavail-ability of lipophilic compounds, as they act as a protective shield, preventing the compounds from degradation *in vivo* (Torchilin 2001, 2008). Several other advan-tages of polymeric micelles include site-specific targeted delivery, a prolonged-release mechanism, low toxicity, and more (Ganta et al. 2008).

4.3 Polymeric Nanogels

Polymer-based nanogels (PNGs) have recently been included in the category of polymeric nanoparticles. These are hydrophilic networks of polymers that are cross-linked through covalent or non-covalent bonds, and they have the ability to absorb water in the aqueous phase. Due to their high moisture content and soft mechanical properties, hydrogels exhibit excellent biocompatibility with tissues and a high drug loading capacity for hydrophilic compounds (Park and Park 1996). Additionally, polymeric nanogels are highly stable due to the presence of covalent bonds within their network. When a stimuli-responsive chain is incorporated into their structure, they can respond to external stimuli such as pH, temperature, chemical agents, and solution properties during drug delivery. The size of polymeric nanogels typically ranges from 20 nm to 500 nm, with an ideal size below 200 nm (Merkel et al. 2011). Ravi et al. (2018) encapsulated the marine carotenoid fucoxanthin using polymeric chitosan-glycolipid nano-carriers to create a nanogel aimed at enhancing the anti-cancer efficacy of fucoxanthin. Interestingly, this led to a 25.8-fold enhancement in caspase-3 activity.

4.4 Dendrimers

Dendrimers are spherical nanoparticles well-known for their unique structure, which consists of hyperbranched monomers originating from a central core and extending outward with peripheral functional groups. The dendrimer structure is created through the polymerization of branching groups using convergent/divergent methods. The resulting structure features a lipophilic core and a hydrophilic surface, making it efficient for encapsulating water-insoluble compounds (Lee et al. 2005).

Dendrimers exhibit various physical and chemical properties, including low viscosity, a macromolecular size, hyperbranched molecular topology, high density, and multiple chemically functionalized groups (Lee et al. 2005). Additionally, the degree of depolymerization can be controlled or tailored to achieve effective control over the release of the loaded compound (Wong et al. 2012).

Chapter 5
Synthetic Polymeric Nanocarriers

5.1 Echogenic Immunoliposomes

Echogenic immunoliposomes are nano-carriers controlled by ultrasound to release encapsulated compounds/drugs from their carriers. These echogenic nano-carriers are fabricated by suspending lipids in an aqueous phase and then adding mannitol, followed by lyophilisation (Alkan-Onyuksel et al. 1996; Huang et al. 2001). The echogenicity of these nano-carriers is due to the presence of entrapped and stabilized gas within the lipid phase during rehydration and sequential lyophilisation (Huang et al. 2002). Echogenic nano-carriers are applicable for drug delivery in specific tissues by modifying their liposomal surface with target-specific antibodies and ligands (Leonetti et al. 1990; Martin et al. 1990). For instance, echogenic nano-carriers attached with antibodies of VCAM-1 are helpful for identifying pathologic endothelium in its initial stage of atherosclerosis growth.

Additionally, the application of ultrasound to disperse the liposomal structure close to the target cell enables an efficient therapeutic effect of the drug on the infected tissue (Hitchcock et al. 2009). In another study, thrombolytic rt-PA was loaded into an echogenic nano-carrier, and its release was triggered by ultrasound in an in vitro model. This loaded nano-carrier is robust and echogenic at 6.9 MHz B-mode when scanned by a clinical diagnostic scanner at a low-pressure output level of MI = 0.04. Furthermore, the concentration of the therapeutic drug rt-PA was released from the nano-carrier at 6.0 MHz using color Doppler ultrasound at an MI of more than 0.43 (Smith et al. 2007).

S. Noore et al., *Nanocarriers for Controlled Release and Target Delivery of Bioactive Compounds*, SpringerBriefs in Food, Health, and Nutrition,
https://doi.org/10.1007/978-3-031-57488-7_5

5.2   Nanobots

The study of robots at the nanoscale (1–100 nm) falls under the domain of nanorobotics within nanotechnology. These nanorobots are capable of performing various biological functions, including signaling, sensing, processing information, and exhibiting swarm behavior at the nanoscale level (Al-Sharif et al. 2010; Neto et al. 2010; Thangavel et al. 2014). In the literature, several synonyms have been cited for nanorobots, such as nonobots, nanites, nano-ids, nano-mites, and more (Rahul 2017; Thangavel et al. 2014). Generally, the term "nanobots"refers to "nano," signifying a small scale level, and "bots," which implies a device that can be controlled or programmed for specific tasks (Rahul 2017).

In the field of medical sciences, nanobots find applications in various categories, including cancer treatments, minimally invasive brain surgery, drug delivery, diabetes monitoring, hemoglobin monitoring, dentistry, spinal cord, and nerve damage treatments (Schmidt et al. 2020; Wang and Zhou 2021; Soto et al. 2020; Rahul 2017; Thangavel et al. 2014; Wang and Zhou 2021). Additionally, nanomedicines have been reported as advanced technology for diagnosing, preventing, and treating diseases, injuries, and pain, thereby contributing to maintaining stable human health (Cavalcanti et al. 2007; Neto et al. 2010; Patel et al. 2006; Thangavel et al. 2014). Among all these applications of nanobots in sciences, the most popular and frequently used in the pharmaceutical and nutraceutical industry is targeted drug delivery to specified sites of infected cells (Upadhyay et al. 2017).

However, nanobots face several significant limitations, including high fabrication costs and challenges in navigating internal interfaces. The high viscosity of blood at the nanoscale level makes it difficult for nanobots to traverse through blood vessels. Additionally, Brownian movement generates collisions among molecules, making their control challenging (Upadhyay et al. 2017). Nanobots must be designed to have a size that does not harm living tissue cells while penetrating the body. They should also be capable of transmitting signals from various sensing systems (Sim and Aida 2017). Addressing these challenges requires further research in the field of nanotechnology in the coming years.

Nanobots typically involve the encapsulation of drugs or bioactive compounds, and their release mechanism depends on physiological factors or external signaling, such as ultrasound triggers (Freitas 2005; Iacovacci et al. 2015). These nanobots are fabricated using polymers, including alginate, silk, biocompatible collagens, poly(glycolic acid), poly(lactic-co-glycolic acid), poly(lactic acid), and poly(caprolactone). Bio nanocomposites have also been developed, incorporating inorganic compounds like bioactive glass, magnetic nanoparticles, silver/gold, carbon and silica nanoparticles, graphene oxide, and titanium oxide (Saha 2009).

In conclusion, nanobots currently lack FDA (US Food and Drug Administration) approval for use within the human body due to associated risks. However, if nanotechnology research advances in the coming years and a better understanding of nanobots'interaction within the human body is achieved, they could become a future-generation diagnostic and drug delivery method, contributing to maintaining organ health with real-time information about each organ's condition.

5.3 Hybrid Nanocarriers

Hybrid nanocarriers encompass two types of nanocarriers: organic-inorganic and lipid-polymer carriers, designed for the controlled release of bioactive compounds. Seyedabadi et al. (2021) developed a slow-release encapsulation method for caffeine using chitosan-coated nano-liposomes, demonstrating superior encapsulation efficiency compared to nano-liposomes without chitosan covering. Thus, the combination of chitosomes proved more effective for caffeine encapsulation. Similarly, Gagliardi et al. (2021) conducted a comparative study involving synthetic and natural-based nanoparticles, specifically poly(lactic-co-glycolic acid) and zein, for rutin encapsulation. The results indicated that zein loaded with 0.8% rutin concentration exhibited a slower release (25%) after 60 min compared to poly(lactic-co-glycolic acid) (100%).

Certain nanostructures, such as cubosomes, tend to degrade in the gastrointestinal tract (GIT) due to the presence of oleic acid secreted by pancreatic lipase and bile salts. To address this issue, surface coating with polyelectrolytes is commonly practiced (Souza et al. 2014). The mechanism behind this coating depends on the charges of the compounds involved, with positively charged polyelectrolytes applied to negatively charged cubosomes. Among natural positively charged polysaccharides, chitosan is the most widely used coating material in the fabrication of lipid-based nanoparticles. Additionally, chitosan exhibits excellent mucoadhesiveness, which enhances the retention time of bioactive compounds (Tan et al. 2015, 2016b).

As reported, chitosan improved the ocular residence and transcorneal permeation of voriconazole encapsulated in cubosomes, resulting in a 171.15% increase in bioavailability compared to normal voriconazole-loaded formulations (Said et al. 2021).

5.4 Decorating Nanocarriers

5.4.1 Quantum Dots

Quantum dots are semiconducting nano-crystals of a cumulative size of 2–10 nm, and they are widely used in detection, cell labeling, and bio imaging (Michalet et al. 2005).

5.4.2   Silver Nanoparticles

Due to their unique physicochemical properties, such as antiviral, antibacterial, anti-angiogenesis, and anticancer effects, silver nanoparticles have been widely utilized in nanomedicine over the years (Gurunathan et al. 2018; Hu et al. 2014; Zhang et al. 2016c). Several methods (physical, chemical, and biological) have been employed to fabricate silver nanoparticles; however, biological methods are considered the best option as they are safe, biocompatible, and eco-friendly. Silver nanoparticles are used in the preparation of medicines for wound dressings, antibacterial lotions for burns, and various skincare products (Rosa et al. 2017).

5.4.3   Gold Nanoparticle

Gold nanoparticles have garnered interest in nanomedicine due to their excellent tolerance of high temperatures. These nanoparticles find applications in cancer studies as well as the detection of DNA sequences, bacteria, viruses, and proteins (Bartczak et al. 2013; Gupta et al. 2016a).

5.4.4   Graphene Oxide

Unique single-atom graphene oxide is a thick two-dimensional hexagonal lattice composed of carbon material. It has gained significant popularity in nanomedicine due to its excellent tolerance to adverse thermal, mechanical, and electrical effects (Ghosal and Sarkar 2018). Owing to its antimicrobial, anticancer, and antiviral properties, it finds extensive application in the treatment of mouth, hand, and foot diseases (Gurunathan et al. 2019).

5.4.5   Zinc Oxide

Zinc oxide is a metallic nanocarrier known for its potent antimicrobial properties, making it a common choice for combatting various microorganisms, including viruses.

Chapter 6
Quality Parameters of Nanoencapsulation

After fabrication of these nanostructures, their characteristics are analyzed based on their structure, morphology, stability, drug loading/releasing efficiency, and molecular attributes. As illustrated in the table, several strategies are employed for analyzing these attributes.

6.1 Particle Size Measurement

The particle size of nano-structures is typically measured using photon correlation spectroscopy (PCS) with a Zetasizer, which operates based on Mie theory. PCS, also known as dynamic or quasi-elastic light scattering, detects fluctuations in scattered light resulting from Brownian motion (Loo et al. 2013). It provides a comprehensive report on the average particle size (z-average) and its polydispersity to help understand a system's particle size distribution. It can characterize particle sizes ranging from nanometers to approximately 3 microns. However, a Laser diffractometer (LD) can measure a broader range of nanoparticles, from nanometers to micrometers. Its mechanism is based on the diffraction pattern that represents the shape and size of the nanoparticles, as illustrated by Fraunhofer theory. Advances in this technology, including a combination of LD and polarization intensity, have enabled the measurement of nanoparticles as small as 10 nm.

Furthermore, the resulting data are encoded by calculating the volume distribution using parameters such as Dv (10), Dv (50), and Dv (90). Dv (90) specifies the size point where up to 90% of the total quantity of particles in a specified sample is accommodated. Similarly, Dv (50) and Dv (10) specify size points where less than 50% and 10% of the sample comprises the measured size, respectively. A span equation is used to estimate particle size distribution in a given sample. Span also calculates the polydispersity index (PI), revealing whether the nanoparticles have a

S. Noore et al., *Nanocarriers for Controlled Release and Target Delivery of Bioactive Compounds*, SpringerBriefs in Food, Health, and Nutrition, https://doi.org/10.1007/978-3-031-57488-7_6

wide or narrow size distribution. Additionally, it assesses particle homogeneity, and its value should fall between 0 and 1. A higher span value indicates a wider particle distribution and greater polydispersity (Patel et al. 2012; Uprit et al. 2013). In theory, a monodisperse particle population reflects a PI value of zero. Therefore, to determine and confirm the particle size distribution, calculating the PI of the given sample is essential (Vickers 2017). The mathematical equation for span calculation is provided below:

$$SPAN = \frac{Dv(90) - Dv(10)}{Dv(50)}$$

Another strategy for measuring particle size is field flow fractionation (FFF). FFF operates based on the principle of particle migration and accumulation and is categorized into three types: sedimentation FFF (gravitational), thermal FFF (thermophoresis/temperature gradient), and Electrical FFF (hydrodynamic fields) (Nastruzzi 2004). In addition to these methods, dynamic ultramicroscopy, ultrasonic spectroscopy, electroacoustic mobility spectroscopy, and cryogenic transmission electron microscopy (cryo-TEM) are also employed to determine the particle size of nanostructures. Particle size determination is essential because it provides detailed information about the shape and size of particles, enabling their application in various fields, such as food formulation and medicine (Babick 2016; Esposito et al. 2015; Rajinikanth and Chellian 2016).

6.2 Zeta Potential

Zeta potential (ZP) plays a crucial role in determining the stability of a nanosuspension. It is calculated based on the electrophoretic mobility of particles in a solution. In general, a high ZP value indicates strong electrical repulsion between particles, reducing the likelihood of particle aggregation, while a low ZP indicates poor stability, often leading to coagulation or flocculation (Lu and Gao 2010). ZP values less than -30 mV or greater than $+30$ mV are typically considered electrostatically stable (Thatipamula et al. 2011).

The determination of ZP is commonly carried out using Laser Doppler electrophoresis with instruments like the Malvern ZetaSizer Nano ZS. An electrical field is applied across the sample, causing oppositely charged particles to migrate towards the electrode at velocities proportional to their zeta potential. These velocities are measured using Laser Doppler velocimetry (Harisa et al. 2017).

The frequency shift of the incident laser beam caused by the migrating particles is used to calculate particle mobility. Subsequently, this data is employed to determine the ZP using Henry's equation. However, certain factors can influence ZP measurements, including the electrical conductivity of the solution, pH value, and the chemical properties of the solution (Chauhan et al. 2020).

6.3 Morphology

The surface morphology of nanostructured particles and suspensions is determined using scanning and transmission electron microscopy (TEM and SEM), atomic force microscopy (AFM), and photon correlation spectroscopy (PCS) (Chauhan et al. 2020). To obtain comprehensive information about these nanoparticles, several sample preparation techniques are employed for TEM, including negative staining, vitrification, and freeze-fracturing of samples. It's important to note that dehydration during the preparation process can lead to structural modifications that alter the actual morphology of the nanoparticles. Cryo-TEM and Cryo-ET (electron tomography) are also utilized to achieve a higher level of visualization (Stewart 2017).

6.4 Entrapment Efficiency

Entrapment or loading efficiency is a ratio between the number of compounds entrapped within the nano-carrier and the total amount of compounds dispersed in the solution during the preparation of nano-particles. This process involves dispersing a known amount of nano-particles in a solution, followed by centrifugation in an ultrafiltration tube. After centrifugation, the supernatant is separated, and the drug concentration is determined using spectrophotometric methods (Stewart 2017). Encapsulation efficiency is then calculated using the following equation:

$$\text{Entrapment efficiency}\,(\%) = \frac{Wa - Ws}{Wa} \times 100\%$$

Where Wa represents the initial mass of the encapsulation compound, and Ws is the concentration level of the compound estimated in the supernatant after centrifugation (Doktorovova and Souto 2009). Loading/entrapment capacity (LC) determines the ratio between the encapsulated compound and the total weight of the wall matrix, and it is calculated using the following formula:

$$LC = \frac{Wa - Ws}{Wa - Ws + W1} \times 100\%$$

Where W1 represents the weight of the wall matrix (Doktorovova and Souto 2009). Additionally, several factors play a significant role in encapsulation efficiency and capacity, including the quality of the surfactant used, the ratio, and the interactions among the loading compound and the wall matrices (Gaba et al. 2015).

6.5 Crystallinity and Polymorphism

Determining the crystallinity and polymorphism of nanoparticles is crucial. These nanoparticles may undergo undesirable polymorphic transitions during storage, leading to the leakage of the encapsulated compound from the wall matrices (Das and Chaudhury 2011). Furthermore, the crystalline structure of nanoparticles also impacts the loading efficiency and release profile of the encapsulated compound (Beck et al. 2011). To identify the crystalline or polymorphic state of the nanoparticles, techniques such as differential scanning calorimetry (DSC), X-ray diffraction (XRD), as well as wide and small-angle X-ray scattering (Grumezescu 2017; Nastruzzi 2004) are employed..

6.6 Magnetic Resonance Investigation

Techniques such as NMR spectroscopy (1H-NMR) and ESR are employed to identify the structure of the entrapped compound and its overall nanostructure (Thompson et al. 2016). Generally, NMR records molecular interactions and the mobility of nanoparticles and encapsulated compounds through signal line widths. The proton relaxation time is associated with the line width of NMR, enabling rapid structural detection. Typically, mobile oil compounds exhibit narrow line widths with greater amplitudes, whereas compounds with limited mobility display low amplitudes and broad signals. Hence, this technique is excellent for detecting immobilization in a sample (Jores et al. 2003; Nastruzzi 2004).

6.7 Raman and Fourier Transform Infrared Spectroscopy

These spectroscopic techniques are typically used to detect interactions between the encapsulated compound and the wall matrices (Shah et al. 2015). Raman spectroscopy, for instance, initiates molecular vibrations using a laser beam, allowing for the mapping of chemical and structural modifications in molecules as they transition from their initial state to a vibrational state. These studies are relevant for gaining insights into the relationships, conformations, and arrangements of molecules within the nanoparticles (Ahmad 2015).

Chapter 7
Nanoencapsulation of Bioactive Compounds

7.1 Bioactive Peptides/Proteins

A protein is composed of more than 50 amino acids (>5000 Da) in the form of a long chain, while peptides consist of a smaller group of less than 50 amino acids (500–5000 Da) linked by peptide bonds. They play various essential roles in the body, including transporting molecules, catalyzing chemical reactions, and responding to stimuli. These molecules are highly susceptible to extreme pH levels and enzymatic degradation, particularly by proteolytic enzymes in the gastrointestinal tract, which break the peptide linkage in proteins and peptides. Additionally, due to their higher molecular weight, proteins cannot effectively cross the intestinal membrane. As a result, extensive research has focused on the oral delivery of these molecules, employing techniques such as chemical modification, PEGylation, peptidomimetics, absorption enhancers, and membrane transporters, often in conjunction with nanoparticle-based delivery systems. Notably, chitosan and solid lipid nanoparticles have been identified as promising nano-carriers for peptide encapsulation (Das et al. 2019). Insulin was encapsulated using cetyl palmitate-based solid lipid nanoparticles, which resulted in a hypoglycemic effect in diabetic rats after 24 h, demonstrating enhanced oral absorption of insulin. Similarly, Reboredo et al. (2021) used zein nano-particles coated with poly(anhydride)-thiamine conjugate to encapsulate insulin-loaded nano-particles, reducing blood glucose levels in rats by up to 20% (Table 7.1).

© The Author(s), under exclusive license to Springer Nature Switzerland AG 2024   49
S. Noore et al., *Nanocarriers for Controlled Release and Target Delivery of Bioactive Compounds*, SpringerBriefs in Food, Health, and Nutrition,
https://doi.org/10.1007/978-3-031-57488-7_7

Table 7.1 Illustration various bioactive encapsulation along with its several encapsulation categories and strategies

Encapsulation category	Wall matrix	Bioactive compounds	Encapsulation strategies	Applications	References
Nanofibers	Hydroxypropyl-β-CD and hydroxypropyl-γ-CD	Cineole and p-cymene	Complex inclusion and electrospinning	Food and oral care applications	Celebioglu et al. (2018)
	Chitosan–gelatin	Thyme EO	Nozzleless electrospinning	Nitrite substitute for meat products	Vafania et al. (2019)
	PVA/β-CD	Cinnamon EO	Electrospinning	Antimicrobial packaging for fresh mushroom	Pan et al. (2019)
	Gelatin	Peppermint and chamomile EO	Electrospinning	Potential application as edible food packaging	Tang et al. (2019)
	Zein	Cinnamic aldehyde	Needleless electrospinning	Food additive to reduce nitrites in sausages	Karim et al. (2021)

Nanocapsule	Chitosan	Pepper tree (Schinus molle) EO	Nano-precipitation	Postharvest control of Colletotrichum gloeosporioides and quality evaluations in avocado	Chávez-Magdaleno et al. (2018)
	Polybutylcyanoacrylate (PBCA)	Green grass fragrance	Emulsion polymerisation	Potential application in food industry	Montanari et al. (2015)
	Polyurethane	EO	Emulsion inversion point	Potential application in food industry	Cui et al. (2018)
	Chitosan	Coriandrum sativum EO	Emulsion formation/ionic gelation	Prolong shelf life and control the fungal and aflatoxin contamination of stored rice	Das et al. (2019)
	Chitosan	Pimpinella anisum EO	Emulsion formation/ionic gelation	Potential safe green food preservative against fungal infestation, aflatoxin B_1	Das et al. (2021)
	Polybutylene adipate-co-terephthalate (PBAT)	Linalool EO	Extrusion	Antimicrobial activity against E. Coli, useful for food packaging	da Silva Barbosa et al. (2021)
	Sorbitan monostearate/poly(ε-caprolactone)	Origanum vulgare/Thymus capitatus	Nano-precipitation	Alternative to synthetic antifungals and potential applications in health and food industry.	Kapustová et al. (2021)
	Methyl- cyclodextrins	Betalains-phenylethylamine-betaxanthin/d indoline-betacyanin	Electrospray ionization	Potential application in functional foods	Matencio et al. (2021)
	Whey protein	Coffee bean oil	Electrospray	Enhanced coffee bean oil efficiency	Reddy and MN)
Nanoemulsion	–	Catechin	Ultrasonication	Food fortification ingredients for food products.	Ruengdech and Siripatrawan (2021)
Core-shell nanofiber	Zein/tragacanth gum	Saffron extract	Coaxial electrospinning	Potential use in food industry (chewing gum and tea bag development)	Dehcheshmeh and Fathi (2019)

(continued)

Table 7.1 (continued)

Encapsulation category	Wall matrix	Bioactive compounds	Encapsulation strategies	Applications	References
Nanoparticle	Polycaprolactone	Geranyl cinnamate	Mini-emulsification/ solvent evaporation technique	Potential use in antimicrobial packaging	Zanetti et al. (2019)
	Chitosan	*Paulownia Tomentosa* EO	Ionic gelation method	Improve shelf-life of ready-to-cook pork chops	Zhang et al. (2019a)
	Zein/carboxymethyl dextrin	Curcumin	Antisolvent precipitation/ evaporation technique	Effective encapsulating materials for bioactive compounds in food and pharmaceutical industry	Meng et al. (2021)
	Zein	*Cinnamodendron dinisii* Schwanke EO	Antisolvent precipitation	Efficient for active packaging specifically in the conservation of ground beef, stabilizing the deterioration reactions and preserving the colour.	Xavier et al. (2021)
	Porcine gelatin	Buriti (*Mauritia flexuosa*) oil	O/W emulsification/ freeze-drying	It enables the water dispersibility and potentiate the antimicrobial activity of Buriti (*Mauritia flexuosa*) oil	Coelho et al. (2021)
	Human serum albumin/L-cysteine	Curcumin	Emulsification	Potential for food and pharmaceutical industries specifically, it improved antitumor activity in vitro	Hao et al. (2020)
	Chitosan/chitosan oligosaccharides/ carboxymethyl chitosan	Cyanidin-3-Oglucoside	Ionic crosslinking via γ-Polyglutamic acid/ calcium chloride	Potential for food and pharmaceutical industries	Sun et al. (2015)
	Tannic acid/Genipin/ human serum albumin	Curcumin	Ionic crosslinking/ ultrasonication/freeze drying	Excellent oral drug delivery for treating ulcerative colitis	Luo et al. (2020)
	Tween 80/soy lecithin	Hydroxycitric acid	Homogenization/ ultrasonication	Improvement in bioavailability of Hydroxycitric acid by two folds	Ezhilarasi et al. (2016)
Nanosponge	CD	Cinnamon EO	Synthesis via a crosslinking agent	Helpful in antimicrobial food packaging	Simionato et al. (2019)

Nanobiocomposite	Chitosan/β-CD citrate/oxidised nano-cellulose	Clove oil	Impregnation of biocomposite	Helpful in active food packaging	Adel et al. (2019)
	Nano-MOFs@CMFP	Curcumin	Impregnation of biocomposite	Food preservation	Toldrá et al. (2020)
	Poly(ethylene-co-vinyl acetate)/chitosan	Iprodione	Nano-precipitation/ultrasonic/freeze drying	The film reflected enhanced antifungal ability and temperature-sensitive drug release	Xiao et al. (2020)
Nanofibrous film	Polylactic acid (PLA)	Thyme EO	Electrospinning	Antimicrobial and humidity sensitive food packaging system	Min et al. (2021)
Nanocomplex	Chitosan hydrochloride/carboxymethyl chitosan/whey protein isolate	Anthocyanins	Complex inclusion	It improve the thermal stability and slow the release of ACNs added to food products	Wang et al. (2021)
Nanoliposomes	Lecithin	Betalains	Ultrasonication	It was incorporated in gummy candies to improve its colour stability and to be explored as a natural colourant for the food industry.	Kumar et al. (2020)
	Chitosan/pectin	Neohesperidin	Ultrasonication	It is a significant nano-carrier for neohesperidin. It enhances the physicochemical stabiligy, controlled drug release and mucoadhesion potential.	Karim et al. (2021)

7.2 Lipids

Lipids, known for their excellent functions such as mobility, absorption, and solubilisation of lipophilic compounds (vitamins A, D, E, and K), help in the secretion of hormones, cellular ingredients, and other necessary compounds that are important for the body's functioning. Essential fatty acids such as polyunsaturated fatty acids (PUFAs) are several carbon atoms attached with two or more double bonds. They are an important part of the cell membrane, and it affects the membrane fluidity and behaviour receptors and enzymes surrounding the membrane (Di Pasquale 2009; Mehrad et al. 2015). As the human body cannot synthesize these essential fatty acids, they need to be taken as a supplement in a diet (Plourde and Cunnane 2007). Due to metabolism deficiency (unable to add a double bond to the fatty acid chain), the human body cannot synthesize these essential fatty acids.

Additionally, these fatty acids (PUFAs) are categorized into four categories based on their molecular structure such as omega-3, omega-6, omega-7, and omega-9. Among all the types of PUFA, the most important one is Omega-9, also called oleic acids, widely available in all fatty food, but its maximum concentration of reported in olive oil. Whereas in the case of omega-6 fatty acids such as linoleic and arachidonic acids are reported to be found majorly in cereals, eggs, vegetable oils, poultry, baked goods, etc. (Das 2006). In addition to this, alpha-linolenic acid (ALA), eicosapentaenoic acid (EPA), and docosahexaenoic acid (DHA) belong to the Omega-3 fatty acids group. Walnut, green vegetables, Canola, flaxseed, linseed, and rapeseed oils contain a high concentration of ALA. At the same time, the major source of EPA and DHA is widely available in fish in cold water. These Omega-three fatty acids help control hypertension, metabolic syndrome X, cancer, atherosclerosis, and collagen vascular diseases (Hosseini et al. 2019). However, during processing, denaturation of these fatty acids into trans-fat develops an adverse effect on health on consumption (Cantwell et al. 2005; Lopez-Garcia et al. 2005). Encapsulation is practiced to reduce the denaturation of these sensitive fatty acids and use them for food formulation and pharmaceutical applications. Post encapsulation, the nano-carriers protect the encapsulated compounds from the adverse effect of the environment, including extreme pH value, oxygen, moisture, high temperature, and light exposure (Esfahani et al. 2019; Yurdugul and Mozafari 2004). Several strategies, such as spray drying, freeze-drying, coacervation, etc., are used to encapsulate essential fatty acids and fish oil (Jafari 2017). Also, nano-liposomes encapsulate these fatty acids to prevent oxidation reactions, thereby improving their bioavailability (Esfahani et al. 2019; Jafari 2017; Ojagh and Hasani 2018). A study reported that omega-3 fatty acid was encapsulated in nano-liposomes and fortified in bread. The results indicated 90% encapsulation efficiency of fish oil in nano-liposome along with improved oxidative stability and no undesirable effect on its sensory attributes (Ojagh and Hasani 2018). Similarly, DHA and EPA were encapsulated in nano-liposome, and their physicochemical attributes were analysed concerning bread and milk, along with their comparison to un-encapsulated fatty acids. The loss in the quality attributes of bread and milk was 4.2–6.5% and 5.6–5.9%, respectively, in the case of encapsulated-fatty acids nano-liposomes. Besides, there was stability during storage. Sensory attributes did not reflect many variations in control and loaded nano-liposomes, but the fishy flavour and aroma were highly detected in un-encapsulated fatty acids (Rasti et al. 2012). In addition, peroxide and anisidine values on encapsulated fatty acids were much lower than in control. According to the authors, this encapsulation method of omega-3 fatty acids for food formulation is highly acceptable and effective (Rasti et al. 2012).

7.3 Minerals

The deficiency of certain minerals in the human body causes several health issues. However, fortification of food products with these minerals can cause undesirable changes in food products such as a change in colour, texture, taste, appearance, etc., and the additional risk of oxidation (Gharibzahedi and Jafari 2017). Presently, iron deficiency is considered one of the major deficiencies among the human race all around the globe. This situation is due to the insufficient intake of iron in the diet or its bioaccessibility (Horton and Ross 2003). Iron deficiency in human blood leads to life-threatening diseases such as anemia, where hemoglobin is reduced below the standard level (Gaucheron 2000). As indicated in literature an antagonism between calcium content of milk and iron, milk is low in iron concentration, and thus infants and young children are iron deficient. Enrichment of milk with iron is important to reduce the iron-calcium antagonism, metallic off-flavour, and consequences of fat oxidation (Ziegler 2011).

Further, to achieve milk fortified with iron, an encapsulated iron in a suitable carrier has been reported as iron sulphate when added directly to food products, causing rapid change in its sensory attributes. When added to iron, Encapsulated iron resulted in no change in its sensory properties for 2 weeks (Zarrabi et al. 2020). Literature indicates fortification of food products with iron can prevent diseases (anemia) caused due to iron deficiency (Simiqueli et al. 2019). However, iron undergoes rapid oxidation; therefore, during the fabrication of an iron nanoencapsulation system as nano-liposomes, antioxidants (vitamin E and ascorbic acids) are included to avoid the oxidation of ferrous ions. In the case of tocosome encapsulation, adding antioxidants is not necessary due to the presence of TP and T2P, which is rich in the antioxidant compound itself (Gianello et al. 2005) (Jafari 2017; Lopez-Garcia et al. 2005). Another important mineral that is generally used in food fortification is magnesium.

Food products fortified with magnesium are widely used in reducing the risk of several lifestyle disorders such as cardiovascular, diabetes (type 2), weakness of muscles, and hypertension (Gharibzahedi and Jafari 2017). In a study, W/O/W emulsion was prepared using rapeseed oil, olein, olive oil, etc., by mixing hydrophobic (polyglycerol polyricinoleate) and hydrophilic (sodium caseinate) emulsifiers to encapsulate magnesium in the aqueous phase (Bonnet et al. 2009). The results indicated magnesium's effective release mechanism based on diffusion and thermal stability. It was also reported that the direct addition of magnesium in milk caused coagulation of milk on heating, while encapsulated magnesium caused no negative effect (Gharibzahedi and Jafari 2017).

7.4   Vitamin E

It is an endogenous antioxidant bioactive compound majorly found in natural oils, providing phenolic hydrogen to free radicals. It is a lipid-soluble antioxidant, extensively used by the food and nutraceutical as an essential compound to inhibit oxidation. It helps in maintaining oxidative stability of frying oils, but this bioactive compound is heat sensitive. Encapsulation helps in protecting the compound and making it stable and highly soluble. Several researchers have used encapsulation strategies to enhance the oxidative property and improve the heat tolerance of *α-tocopherol*. Ma et al. (2021) encapsulated *α-tocopherol* using ethyl cellulose (M200) in soybean oil. They observed the effective oxidative stability of soybean oil during thermal processing by reducing the total polar count (44.29%) and carbonyl value (46.94%). Further, Singh et al. (2018) encapsulated vitamin E using sodium alginate (1.5%) and pectin (2%), which enhanced its oxidative stability with practical applications in enriching fat-based bakery products to prevent autoxidation. Similarly, Sahafi et al. (2021) encapsulated *α-tocopherol* in pomegranate seed oil which enhanced the thermal stability (20–90 °C) and extended the shelf life from 20 to 50 days.

7.5   Vitamin C

A water-soluble compound naturally found in tomatoes, berries, citrus fruits, and potatoes is majorly utilized by nutraceutical, pharmaceutical, and food industries as an antioxidant bioactive compound. It helps in several physiological functions of the body, including hydroxylation reactions in collagen synthesis, growth, and repair of connective tissues and skin. However, when exposed to heat, light, oxygen, and pH changes, its unstable behaviour rapidly gets degraded to dehydroascorbic acid and hydrolysed to form 2,3-L-diketogulonic. Hence, encapsulation helps enhance the stability of this bioactive compound by delaying its degradation. Baek et al. (2021) encapsulated vitamin C using cellulose and chitosan. The encapsulation resulted in the controlled release from 6 to 8 h and enhanced antioxidant and antimicrobial properties. The microbial count for *Escherichia coli and Staphylococcus aureus* were in the range of 8 and 16 µg/mL, which proved to be efficient for extending the shelf -life of the food products. Similarly, Yan et al. (2021) developed vitamin C-rich gummy candies after encapsulating them in casein gel, which enhanced retention of vitamin C (92%) after a period of 10-weeks compared to conventional gummies with only 79% retention.

7.6   Plant Extracts and Essential Oils

Essential Oils (Eos) such as carvacrol, thymol, eugenol, linalool, and cinnamalde-hyde combine several bioactive compounds, including acids, esters, ketones, alcohols, and aldehydes groups, which exhibit significant antioxidant and antimicrobial activities. These EOs are widely used as a bioactive ingredient by cosmetics, bio-pesticide, food packaging, and preservation industries (Froiio et al. 2019). Zhang et al. (2019a) developed an edible coating by encapsulating *Paulownia tomentosa* (PT) EO in a chitosan solution. The result reveals that the ready-to-eat pork chops' shelf life increases from 7 to 16 days by incorporating PT-EO-based chitosan film. A similar experiment was conducted by Xavier et al. (2021) using *Cinnamodendron dinisii* EO loaded in zein to prepare chitosan-based packaging film. The study reveals that this film enhanced the shelf life of fresh meat from 4 to 12 days in refrigerated conditions. In another study, Emamjomeh et al. (Emamjomeh et al. 2021) prepared a nano-emulsion by encapsulating *Zataria multiflora* and *Eucalyptus globulus* using maltodextrin (35% w/w) and angum gum (5% w/w). The prepared emulsion enhanced the toxicity effect of these EOs from 4.33 to 10.74 times against instar larvae of *Ephestia kuehniella* at a lethal concentration of 90%.

Extract from plants including Basil (*Ocimum basilicum*), Neem oil (*Azadirachta indica*), Ajowan oil (*Trachyspermum Ammi L*), and green tea extract (*Camellia sinensis var Assamica*) contain major bioactive compounds including phenolic compounds, phenolic diterpenes, tannins, sulfur-containing compounds, alkaloids, and vitamins. These extracts are widely used in antifungal, antibacterial, antispasmodic, anti-germicidal, and anti-fungicidal characteristics (Shahidi and Hossain 2018).

However, to protect these plant extracts' physical, morphological, and sensory attributes, several biobased nano-carriers, including pectin, inulin, and alginate, are employed to encapsulate these bioactive compounds. These are either sprayed or applied as a functional layer on the inner side of the packaging material or incorporated directly into food products. Belscak-Cvitanovic et al. (2015) encapsulated green tea extract in oligosaccharide. They added the encapsulated powder in chocolate drinks as a functional ingredient to enhance the nutritional profile of the drink. Further, Taylor et al. (2007) encapsulated nisin using di-stearoyl phosphatidylcholine and distearoyl phosphatidylglycerol to form liposomes. Results revealed that the encapsulated nisin was resistant in a temperature range of 25–75 °C and pH ranging from 5.5 to 11.0.

7.7   Natural Pigments

7.7.1   *Curcumin*

The most extensively utilized compound for nanoencapsulation is curcumin. Curcumin is the major compound in *Curcuma longa* rhizome which is utilized for its colour, antioxidant, antifungal, and health benefits (Brahatheeswaran et al. 2012).

However, its application is severely affected due to poor oral bioavailability, chemical solubility, and low water solubility. Therefore, several encapsulation strategies are employed to improve the effectiveness of this bioactive compound. Pathak et al. (2015) encapsulated curcumin using chitosan. They found that the anti-lipid peroxidation activity of encapsulated curcumin enhanced from 40% to 50%, whereas the shelf life was extended to 15 days at 4 °C. Further, Zhang et al. (2020) encapsulated curcumin using myosin at 40–80 µg/mL concentration. The results reveal that about 80% thermal resistance was achieved with enhanced chemical stability and improved aqueous solubility (Zhang et al. 2020). In a further study, Zhang et al. (2021) developed fucoidan stabilized zein nano-particles to encapsulate curcumin, and this nano-emulsion was thermal (80 °C) and pH (2–8) resistant with a shelf-life of 28 days at 4 °C.

7.7.2   Anthocyanins

Anthocyanins are pigments with excellent antioxidant activity, useful for treating cardiovascular diseases, diabetes, and cancer (Sun et al. 2016). According to the reports, soybean protein nano-cage (H-2 ferritin) was employed to encapsulate cyanidin-3-O-glucoside (anthocyanin). Results indicated that encapsulated cyanidin-3-O-glucoside passed efficiently through the cell compared to non-encapsulated. Additionally, the nano-cage of ferritin was extremely tolerant to high temp and physicochemical denaturants (Zhang et al. 2014).

7.7.3   Carotenoids

It is commonly recognized as a source of fat-soluble vitamins and comes in various colors ranging from red to yellow. Carotenoids play a significant role in human health by shielding cells and tissues from damage caused by free radicals, singlet oxygen, and UV radiation. They also bolster the immune system and inhibit the growth of cancer cells. Notable carotenoids utilized by the food processing industry include lutein, lycopene, zeaxanthin, astaxanthin, β-cryptoxanthin, α-carotene, and β-carotene. These compounds are abundant in fruits like carrots, tomatoes, peppers, pumpkins, and dark green leafy vegetables, often concealed by their rich chlorophyll content. Citrus waste, such as orange and lemon peel, also contains substantial amounts of carotenoids (Xavier et al. 2021). However, due to numerous double bonds, it has poor solubility in water which is why nanoencapsulation is applied to these compounds to enhance their bioavailability and improve their solubility. Kim et al. (2010) encapsulated astaxanthin using β-cyclodextrins which improved water solubility of astaxanthin by 110-fold and thermal resistance by 100 °C. It also enhanced the stability of astaxanthin by 7–9 folds compared to conventional astaxanthin. Further, Savic Gajic et al. (2021) encapsulated carotenoids in calcium alginate beads to improve their stability. This encapsulated carotenoid was used to develop carotene-rich olive oil. The result reveals a negligible change in the oxidation activity of carotene-rich oil compared with non-encapsulated carotenoids.

7.8   Flavours and Aroma

It has always been a challenge to encapsulate flavour and aroma in nano-technology. Presently, nanoencapsulation of these volatile compounds is limited due to certain drawbacks such as high cost of production, stability of the encapsulated volatile compound, shelf life, etc. (Best 2000). The technology can be widespread if its efficacy and cost of production could be reduced; therefore, much extensive research is still needed in this area (Ghasemi et al. 2018; Zuidam and Heinrich 2010). Aromatic compounds are generally composed of several functional groups such as alcohols, aldehydes, ketones, esters, and small molecular weight acids (100–250 Da) (Trifković et al. 2016). Nanoparticles such as nanoliposomes and tocosomes encapsulate aromatic and flavouring compounds in the aqueous phase. This type of nano-structures also encapsulates lipophilic aromatic and flavour compounds in the aqueous phase. Alteration in the bilayer of the liposome composition by the adjustment of phase transition temperature (Tc), thereby enabling these compounds to remain encapsulated and protected from degradation during the storage. However, compounds are released on consumption or by re-heating the product into which it has been incorporated (Arnaud 1995). One of the researchers reported that volatile compounds are released during microwave treatment. The presence of phospholipid bi-layer produces a release of more desirable flavour in low-fat food products (van Nieuwenhuyzen and Szuhaj 1998). Therefore, the encapsulation of volatile flavours in nanoliposomes with phospholipid by-layer will enable efficient encapsulation and the stimulus release of the entrapped compounds, thereby providing customer satisfaction with the end product. This stimulus action is generally triggered during the heating process, and the release of these compounds lasts till the time of ingestion (Kaddah et al. 2018; Peschka et al. 1998).

7.9   Non-anthocyanin Phenolic Compounds

Majorly phenolic compounds are found in fresh produce, including strawberries, apples, pear, grapes, lettuce, and tomatoes. Quercetin has higher antioxidant activity than ascorbyl, rutin, and Trolox due to the position of the hydroxyl group in its chemical structure (Piskula et al. 2012). It possesses health benefits such as antiviral, anti-cancer, and anti-inflammatory. It cures cardiovascular diseases; however, due to its low chemical instability, short biological half-life, low bioavailability, and less water solubility reduce its efficiency when incorporated into food, nutraceutical, and pharmaceutical products. To overcome these barriers, several studies are carried out to enhance the stability and solubility of quercetin. Marcela et al. developed quercetin-loaded lecithin-chitosan nanoparticles, and the results reveal that the stability of quercetin was enhanced from 6 to 28 days at 4 °C. Similarly, Patil et al. (Patil and Killedar 2021) used chitosan, glyceryl monooleate, and poloxamer 407 to develop quercetin-loaded nanoparticles. Results reveal that quercetin-loaded nanoparticles have better quercetin delivery at colonic pH than conventional quercetin. A similar line of the experiment was carried out by Chavoshpour-Natanzi (Chavoshpour-Natanzi and Sahihi 2019), where β-lactoglobulin was used as a wall matrix to entrap quercetin. β-lactoglobulin being resistant to peptide digestion resulted in an enhanced delivery rate compared to untreated quercetin.

7.10 Bacteriophages

Bacteriophages (phages) are viruses that infect specific bacteria. Lytic phages have many advantages as biological control agents: (i) they are highly abundant in nature and their isolation and production inexpensive; (ii) they multiply after infection, persisting in the environment as long as there exist target bacteria; (iii) they can adapt to the bacterial defence mechanisms. In a global context of increasing antibiotic resistance and concern about the impact of microbiota on health, research is increasing into the use of bacteriophages for preventive and therapeutic purposes and in technological applications to reduce the burden of pathogens in livestock and processed foods. Phages have, however, limited stability in solution. They are sensitive to extreme temperatures and pH values, proteases and nucleases, chelating agents, shear stress, and exposure to UV light, so phages are significantly inactivated during processing and storage. In vitro studies and clinical trials of phage therapy in humans and animals show that the effectiveness of phages depends on their dose, ability to reach the site where the bacteria are multiplying, and the timing of administration, sometimes requiring multiple treatments.

Encapsulation of phages can significantly improve their applicability, increasing stability against external inactivating factors and processing, allowing for targeted release and enabling sustained and/or externally stimulated release approaches (e.g. encapsulation in pH-sensitive particles) that may reduce the need for multiple treatments. For these purposes, phages have been encapsulated using synthetic and natural polymers by techniques such as freeze-drying, spray-drying, or emulsification with solvent evaporation. For instance, Salmonella Enteritidis phage f3αSE encapsulated into alginate beads produced by extrusion (Soto et al. 2020) increased the resistance to pH and temperature (0% viability for the free phage vs 80% for the encapsulated phage after 1 h at pH 3, and + 100 h resistance compared to the free phage in a water flow system mimicking the poultry processing industry). Salmonella-specific phage Felix O1 micro-encapsulated into chitosan-alginate beads produced using extrusion and ionic gelation resisted simulated gastric digestion and maintained the viability after 3 h of incubation in 1% and 2% bile solutions, in which the free phage counts decreased by 1.29 and 1.67 log units, respectively. Also, the encapsulated phage retained 100% viability when stored at 4 °C for 6 weeks. Synthetic polymers were also used to encapsulate phage Felix O1; this phage was loaded into pH-responsive particles by spray-drying using Eudragit S100, a poly-methacrylate polymer that dissolves only above pH 7. The resulting beads improved the survivability of phages to gastric acidity and allowed the release of the phages in conditions mimicking the intestinal environment.

7.11   Other Small Molecules

Over the years, several drugs such as cisplatin, siRNA, and doxorubicin have been nanoencapsulated within protein ferritin nanocages for their target delivery on a specified site (Andreata et al. 2020; Huang et al. 2020; Nasrollahi et al. 2020).

Chapter 8
Controlled Release and Target Delivery of Nanoencapsulated Compounds

Based on the application of the encapsulated compounds, its release mechanism is tailored, such as a rapid-release system in local anesthetic lotions and slow release in sunscreens lotions. The release is generally controlled by the structure of the carrier matrices and Ficks's diffusion, swelling, erosion, etc., mechanism. For example, if the compound is loaded in the outer shell of the nanocarrier, the release profile is rapid such as in the case of bioactive cyclosporine (Muller and Runge 1998). The release is slow if the compound is mixed in carrier matrice in solid-state (molecular dispersion). If the compound is entrapped within the core of the carrier matrix, the release is relatively slow. The matrix of the nanocarriers is controlled by its composition and physicochemical properties, such as solubility in different media. Once the nanosystem interacts with the changing environment, such as a change in pH value, temperature, ionic strength, etc., the structure tends to shift based on the above mentions mechanism. These mechanisms are broadly categorized into three different classes, including class I ideal fickian diffusion (Brownian transport), class II polymer relaxation-driven transport, class III anomalous behaviour (combination of class I and class II) (Bourbon et al. 2016a). During the preparation of nanostructures, if the concentration of the bioactive compound is soluble in liquid lipid but insoluble in a solid lipid, the compound will precipitate on cooling resulting in an active core coated with some dissolved compounds in lipid. But if the lipid solidifies rapidly, a bioactive shell will be formed. Generally, the prolonged-release is applicable for encapsulation of bioactive compounds as it tends to degrade in GIT due to the lipase-colipase complex. Therefore, the compounds' release mechanism needs to be understood before fabricating nanostructures for its effective application in food formulations, pharmaceuticals, and nutraceuticals. Several experiments on the release mechanism of encapsulated compounds have been conducted in recent years (Bourbon et al. 2016b, 2003; Ramos et al. 2016). In one of the studies, the effect of GIT condition and release mechanism of curcumin and caffeine was evaluated in protein-based nano-hydrogels coated with chitosan. The results

S. Noore et al., *Nanocarriers for Controlled Release and Target Delivery of Bioactive Compounds*, SpringerBriefs in Food, Health, and Nutrition, https://doi.org/10.1007/978-3-031-57488-7_8

indicated enhanced stability and bioaccessibility of the bioactive compound *in vitro* GIT conditions (Bourbon et al. 2018). Stimuli-sensitive nanocarriers tend to modify their physicochemical properties and structure based on the internal or external stimulus actions. Presently, stimuli-based delivery of bioactive compounds with time and site-specific has gained popularity in the nutraceutical and pharmaceutical industries. Several modes of stimuli mechanisms have been developed, such as chemical (change in pH value), biological (wall matrices degradation enzyme), and physical (light, ultrasound, and temperature) for effective controlled target delivery of encapsulated bioactive compounds/drugs. Additionally, the stimulus release is extremely beneficial in the case of disease management, as when a unique stimulus compound is fabricated in nanocarriers, it specifically responds to the pathological trigger of that stimulus (Daglar et al. 2014).

8.1   Commercial Products

Nanostructured particles reflect exquisite physicochemical properties. However, due to their nanosize, many unseen effects of nanostructures on the human body need to be analysed, for example, its accumulation on different parts of the body/ organs. Therefore prior to application and mass fabrication of these nanostructures, in vitro, in vivo, and silico tries are used to assess its effect and further application (Katouzian and Jafari 2016). The first nanoencapsulated product launched in the market was the cosmetic nano repair product in 2005 by Dr. Rimpler GmbH Germany (http://www.rimpler.de). In the following year, a South Korean company called AmorePacific launched its first cosmetic skin product called 'IOPE.' Over the years, several products have been launched employing nanotechnology (Pardeike et al. 2009). Nanotechnology-based products improved efficacy, such as reducing wrinkles, skin hydration, and diffusion of active compounds in skin tissues. In addition, worldwide distributors for lipid-based nanocarriers are Berg + Schmidt company in Hamburg, Germany (http://www.berg-schmidt.de/). They have launched lipid-based nanoparticles with a trading name called 'BergaCare smartLipids' in the cosmetic market. Presently these lipid-based nanoparticles are available with encapsulated coenzyme Q 10 and retinol. These encapsulated products are sold at 10–20% concentration levels. Besides, customized products are also available, customers are required to provide a specific requirement, and the products will be fabricated based on their preferences and formulations (Pyo et al. 2017). In addition to this, several lipid-based nanoproducts are available in the cosmetic and pharmaceutical market; however, in the food industry, only a few products have been launched, such as coatsome™ (Peters et al. 2016), Aquanova NovaSol® (a nanoemulsion encapsulation lipophilic vitamins and ω-3 fatty acids), Fabuless® (a nanoemulsion to slow down digestion till it reaches small intestine's lower part) (McClements 2015), Canola active oil® launched by Shemen industries in Israel (self-assembled liquid lipids for

phytosterols retention), Asia Food beverage® in Thailand (fabrication nanoliposomes for food applications) and Curcosome® (Quercetin-loaded nanoliposomes).
The further latest information on nanomaterial for food applications is available at
http://www.nanotechproject.org.

Chapter 9
Future Challenges and Conclusions

The growing volume of reports and research publications related to the development of nano-based structures highlights the significant interest demonstrated by scientists in the food, nutraceutical, and pharmaceutical fields. Consequently, this growing interest has ushered in fresh opportunities for the application of nanostructures in the creation of advanced future products within these industries. The adoption of nanostructures for encapsulating drugs and bioactive compounds, thereby enabling controlled and targeted release, has gained global recognition. Nevertheless, significant challenges persist, including the need to reduce production costs on an industrial scale. Furthermore, addressing issues such as enhancing the gastrointestinal stability of protein-based nanocarriers and conducting comprehensive toxicological studies to validate their safety in human interactions remains imperative.

In summary, it is anticipated that by 2030, the world will witness the widespread integration of encapsulated bioactive compounds and drugs into food formulations, nutraceuticals, and pharmaceutical products. These innovations will feature elevated levels of nutrients and antioxidants, accompanied by extended shelf life, improved stability and bioaccessibility. Additionally, the implementation of nano-encapsulation strategies holds the promise of mitigating the most prevalent food degradation phenomenon—the browning reaction.

S. Noore et al., *Nanocarriers for Controlled Release and Target Delivery of Bioactive Compounds*, SpringerBriefs in Food, Health, and Nutrition, https://doi.org/10.1007/978-3-031-57488-7_9

Acknowledgement

The authors would like to place sincere acknowledgments to Food Processing and Nutrition, International Iberian Nanotechnology Laboratory, Braga, Portugal for providing opportunity in their laboratory to work on nanoencapsulation of bioactive compounds.

© The Editor(s) (if applicable) and The Author(s), under exclusive license to
Springer Nature Switzerland AG 2024
S. Noore et al., *Nanocarriers for Controlled Release and Target Delivery of Bioactive Compounds*, SpringerBriefs in Food, Health, and Nutrition,
https://doi.org/10.1007/978-3-031-57488-7

References

Abaee A, Mohammadian M, Jafari SM (2017) Whey and soy protein-based hydrogels and nano-hydrogels as bioactive delivery systems. Trends Food Sci Technol 70:69–81

Abbasi A et al (2014) Stability of vitamin D3 encapsulated in nanoparticles of whey protein isolate. Food Chem 143:379–383

Adamcik J et al (2010) Understanding amyloid aggregation by statistical analysis of atomic force microscopy images. Nat Nanotechnol 5(6):423–428

Adel AM et al (2019) Inclusion complex of clove oil with chitosan/β-cyclodextrin citrate/oxidized nanocellulose biocomposite for active food packaging. Food Packag Shelf Life 20:100307

Ahmad MU (2015) Lipids in nanotechnology. Elsevier

Akita F et al (2007) The crystal structure of a virus-like particle from the hyperthermophilic archaeon Pyrococcus furiosus provides insight into the evolution of viruses. J Mol Biol 368(5):1469–1483

Akiyama Y et al (2007) Preparation of stimuli-responsive protein nanogel by quantum-ray irradiation. Colloid Polym Sci 285(7):801–807

Akkermans C et al (2008) Peptides are building blocks of heat-induced fibrillar protein aggregates of β-lactoglobulin formed at pH 2. Biomacromolecules 9(5):1474–1479

Albertsson A-C, Varma IK (2002) Degradable aliphatic polyesters, vol 157. Springer

Ali A et al (2022) Effect of co-encapsulated natural antioxidants with modified starch on the oxidative stability of β-carotene loaded within Nanoemulsions. Appl Sci 12(3):1070

Alkan-Onyuksel H et al (1996) Development of inherently echogenic liposomes as an ultrasonic contrast agent. J Pharm Sci 85(5):486–490

Al-Sharif M, Awen B, Molvi K (2010) Nanotechnology in cancer therapy: a review. J Chem 2(5):161–168

Andreata F et al (2020) Co-administration of H-ferritin-doxorubicin and Trastuzumab in neoadjuvant setting improves efficacy and prevents cardiotoxicity in HER2+ murine breast cancer model. Sci Rep 10(1):1–10

Aniesrani Delfiya DS, Thangavel K, Amirtham D (2016) Preparation of curcumin loaded egg albumin nanoparticles using acetone and optimization of desolvation process. Protein J 35(2):124–135

Arnaud JP (1995) Pro-liposomes for the food industry. Food Technol Europe 2:30–34

Arpagaus C et al (2017) Nanocapsules formation by nano spray drying. In: Nanoencapsulation technologies for the food and nutraceutical industries. Elsevier, pp 346–401

S. Noore et al., *Nanocarriers for Controlled Release and Target Delivery of Bioactive Compounds*, SpringerBriefs in Food, Health, and Nutrition,
https://doi.org/10.1007/978-3-031-57488-7

Aslan B et al (2013) Nanotechnology in cancer therapy. J Drug Target 21(10):904–913

Assadpour E et al (2016a) Evaluation of folic acid nano-encapsulation by double emulsions. Food Bioprocess Technol 9(12):2024–2032

Assadpour E et al (2016b) Optimization of folic acid nano-emulsification and encapsulation by maltodextrin-whey protein double emulsions. Int J Biol Macromol 86:197–207

Azzi A (2006) Tocopheryl phosphate, a novel natural form of vitamin E: In vitro and in vivo studies: Wiley Online Library

Babick F (2016) Suspensions of colloidal particles and aggregates, vol 20. Springer

Baek J et al (2021) Encapsulation and controlled release of vitamin C in modified cellulose nanocrystal/chitosan nanocapsules. Curr Res Food Sci 4:215–223

Bartczak D et al (2013) Manipulation of in vitro angiogenesis using peptide-coated gold nanoparticles. ACS Nano 7(6):5628–5636

Bastos LPH et al (2020) Encapsulation of black pepper (Piper nigrum L.) essential oil with gelatin and sodium alginate by complex coacervation. Food Hydrocoll 102:105605

Beck R, Guterres S, Pohlmann A (2011) Nanocosmetics and nanomedicines: new approaches for skin care. Springer

Belščak-Cvitanović A et al (2015) Efficiency assessment of natural biopolymers as encapsulants of green tea (Camellia sinensis L.) bioactive compounds by spray drying. Food Bioprocess Technol 8(12):2444–2460

Benshitrit RC et al (2012) Development of oral food-grade delivery systems: current knowledge and future challenges. Food Funct 3(1):10–21

Best D (2000) Ingredient trends alert. Food Process 2000:57–62

Bolder SG et al (2007) Effect of stirring and seeding on whey protein fibril formation. J Agric Food Chem 55(14):5661–5669

Bonnet M et al (2009) Release rate profiles of magnesium from multiple W/O/W emulsions. Food Hydrocoll 23(1):92–101

Bourbon AI et al (2011) Physico-chemical characterization of chitosan-based edible films incorporating bioactive compounds of different molecular weight. J Food Eng 106(2):111–118

Bourbon AI et al (2015) Development and characterization of lactoferrin-GMP nanohydrogels: evaluation of pH, ionic strength and temperature effect. Food Hydrocoll 48:292–300

Bourbon AI, Cerqueira MA, Vicente AA (2016a) Encapsulation and controlled release of bioactive compounds in lactoferrin-glycomacropeptide nanohydrogels: curcumin and caffeine as model compounds. J Food Eng 180:110–119

Bourbon AI et al (2016b) Influence of chitosan coating on protein-based nanohydrogels properties and in vitro gastric digestibility. Food Hydrocoll 60:109–118

Bourbon AI et al (2018) In vitro digestion of lactoferrin-glycomacropeptide nanohydrogels incorporating bioactive compounds: effect of a chitosan coating. Food Hydrocoll 84:267–275

Brahatheeswaran D et al (2012) Hybrid fluorescent curcumin loaded zein electrospun nanofibrous scaffold for biomedical applications. Biomed Mater 7(4):045001

Braich AK et al (2022) Amla essential oil-based nano-coatings of Amla fruit: analysis of morphological, physiochemical, enzymatic parameters, and shelf-life extension. J Food Proc Preservat 46:e16498

Buonocore GG et al (2003) Modeling the lysozyme release kinetics from antimicrobial films intended for food packaging applications. J Food Sci 68(4):1365–1370

Cantwell MM et al (2005) Contribution of foods to trans unsaturated fatty acid intake in a group of Irish adults. J Hum Nutr Diet 18(5):377–385

Carrillo-Conde B et al (2011) Mannose-functionalized "pathogen-like" polyanhydride nanoparticles target C-type lectin receptors on dendritic cells. Mol Pharm 8(5):1877–1886

Cavalcanti A et al (2007) Nanorobot architecture for medical target identification. Nanotechnology 19(1):015103

Celebioglu A, Yildiz ZI, Uyar T (2018) Electrospun nanofibers from cyclodextrin inclusion complexes with cineole and p-cymene: enhanced water solubility and thermal stability. Int J Food Sci Technol 53(1):112–120

Chaudhari AK et al (2021) Nanoencapsulation of essential oils and their bioactive constituents: a novel strategy to control mycotoxin contamination in food system. Food Chem Toxicol 149:112019

Chauhan I et al (2020) Nanostructured lipid carriers: a groundbreaking approach for transdermal drug delivery. Adv Pharm Bull 10(2):150–165

Chávez-Magdaleno ME et al (2018) Effect of pepper tree (Schinus molle) essential oil-loaded chitosan bio-nanocomposites on postharvest control of Colletotrichum gloeosporioides and quality evaluations in avocado (Persea americana) cv Hass. Food Sci Biotechnol 27(6):1871–1875

Chavoshpour-Natanzi Z, Sahihi M (2019) Encapsulation of quercetin-loaded β-lactoglobulin for drug delivery using modified anti-solvent method. Food Hydrocoll 96:493–502

Chen H, Li M-H (2021) Recent progress in polymer cubosomes and hexosomes. Macromol Rapid Commun 42(15):2100194

Chen L, Remondetto GE, Subirade M (2006) Food protein-based materials as nutraceutical delivery systems. Trends Food Sci Technol 17(5):272–283

Chen N et al (2014) Stable and pH-sensitive protein nanogels made by self-assembly of heat denatured soy protein. J Agric Food Chem 62(39):9553–9561

Chen F-P, Li B-S, Tang C-H (2015) Nanocomplexation of soy protein isolate with curcumin: influence of ultrasonic treatment. Food Res Int 75:157–165

Chen H et al (2016) Engineering protein interfaces yields ferritin disassembly and reassembly under benign experimental conditions. Chem Commun 52(46):7402–7405

Chen H et al (2021) The development of natural and designed protein nanocages for encapsulation and delivery of active compounds. Food Hydrocoll 121:107004

Cheng X et al (2017) Folic acid-modified soy protein nanoparticles for enhanced targeting and inhibitory. Mater Sci Eng C 71:298–307

Chinnaiyan SK et al (2022) Fabrication of basil oil nanoemulsion loaded gellan gum hydrogel—evaluation of its antibacterial and anti-biofilm potential. J Drug Deliv Sci Technol 68:103129

Coelho SC, Estevinho BN, Rocha F (2021) Encapsulation in food industry with emerging electrohydrodynamic techniques: electrospinning and electrospraying–a review. Food Chem 339:127850

Cruz-Romero MC et al (2013) Antimicrobial activity of chitosan, organic acids and nano-sized solubilisates for potential use in smart antimicrobially-active packaging for potential food applications. Food Control 34(2):393–397

Cui G et al (2018) Preparation and properties of narrowly dispersed polyurethane nanocapsules containing essential oil via phase inversion emulsification. J Agric Food Chem 66(41):10799–10807

da Silva Barbosa RF, de Souza AG, Rangari V, dos Santos Rosa D (2021) The influence of PBAT content in the nanocapsules preparation and its effect in essential oils release. Food Chem 344:128611

Daglar B et al (2014) Polymeric nanocarriers for expected nanomedicine: current challenges and future prospects. RSC Adv 4(89):48639–48659

Das UN (2006) Essential fatty acids-a review. Curr Pharm Biotechnol 7(6):467–482

Das S, Chaudhury A (2011) Recent advances in lipid nanoparticle formulations with solid matrix for oral drug delivery. AAPS PharmSciTech 12(1):62–76

Das S et al (2019) Encapsulation in chitosan-based nanomatrix as an efficient green technology to boost the antimicrobial, antioxidant and in situ efficacy of coriandrum sativum essential oil. Int J Biol Macromol 133:294–305

Das S et al (2021) Nanostructured Pimpinella anisum essential oil as novel green food preservative against fungal infestation, aflatoxin B1 contamination and deterioration of nutritional qualities. Food Chem 344:128574

de Souza ML et al (2022) Wild Passiflora (Passiflora spp.) seed oils and their nanoemulsions induce proliferation in HaCaT keratinocytes cells. J Drug Deliv Sci and Technol 67:102803

Dehcheshmeh MA, Fathi M (2019) Production of core-shell nanofibers from zein and tragacanth for encapsulation of saffron extract. Int J Biol Macromol 122:272–279

Demirci M et al (2017) 3-encapsulation by nanoliposomes A2-Jafari, Seid Mahdi. In: Nanoencapsulation technologies for the food and nutraceutical industries. Academic, pp 74–113

Di Pasquale MG (2009) The essentials of essential fatty acids. J Dietary Suppl 6(2):143–161

Doktorovova S, Souto EB (2009) Nanostructured lipid carrier-based hydrogel formulations for drug delivery: a comprehensive review. Expert Opin Drug Deliv 6(2):165–176

Donato-Capel L et al (2014) Technological means to modulate food digestion and physiological response. In: Food structures, digestion and health. Elsevier, pp 389–422

Dong F et al (2016) Doxorubicin-loaded biodegradable self-assembly zein nanoparticle and its anti-cancer effect: preparation, in vitro evaluation, and cellular uptake. Colloids Surf B: Biointerfaces 140:324–331

Đorđević TM, Đurović-Pejčev RD (2015) Dissipation of chlorpyrifos-methyl by Saccharomyces cerevisiae during wheat fermentation. LWT-Food Sci Technol 61(2):516–523

Dubrez L et al (2020) Heat-shock proteins: chaperoning DNA repair. Oncogene 39(3):516–529

Ebrahimi HA et al (2015) Repaglinide-loaded solid lipid nanoparticles: effect of using different surfactants/stabilizers on physicochemical properties of nanoparticles. DARU J Pharm Sci 23(1):1–11

Ekwall B et al (1990) Toxicity tests with mammalian cell cultures. In: Short-term toxicity tests for non-genotoxic effects. Wiley, New York, pp 75–97

Emamjomeh L et al (2021) Nanoencapsulation enhances the contact toxicity of Eucalyptus globulus Labill and Zataria multiflora Boiss essential oils against the third instar larvae of Ephestia kuehniella (Lepidoptera: Pyralidae). Int J Pest Manag 69:1–9

Esfahani R et al (2019) Loading of fish oil into nanocarriers prepared through gelatin-gum Arabic complexation. Food Hydrocoll 90:291–298

Esfanjani AF et al (2015) Nano-encapsulation of saffron extract through double-layered multiple emulsions of pectin and whey protein concentrate. J Food Eng 165:149–155

Esfanjani AF, Jafari SM, Assadpour E (2017) Preparation of a multiple emulsion based on pectin-whey protein complex for encapsulation of saffron extract nanodroplets. Food Chem 221:1962–1969

Esmaeilzadeh P et al (2013) Synthesis and characterization of various protein α-lactalbumin nanotubes structures by chemical hydrolysis method. Adv Nanopart 2:154–164

Esmaili M et al (2011) Beta casein-micelle as a nano vehicle for solubility enhancement of curcumin; food industry application. LWT-Food Sci Technol 44(10):2166–2172

Esposito E et al (2015) Production, physico-chemical characterization and biodistribution studies of lipid nanoparticles. J Nanom Nanotechnol 6(1):1

Ezhilarasi PN et al (2013) Nanoencapsulation techniques for food bioactive components: a review. Food Bioprocess Technol 6(3):628–647

Ezhilarasi PN, Muthukumar SP, Anandharamakrishnan C (2016) Solid lipid nanoparticle enhances bioavailability of hydroxycitric acid compared to a microparticle delivery system. RSC Adv 6(59):53784–53793

Fan B et al (2021) Triggered degradable colloidal particles with ordered inverse bicontinuous cubic and hexagonal mesophases. ACS Nano 15(3):4688–4698

Fang J-Y et al (2008) Lipid nanoparticles as vehicles for topical psoralen delivery: solid lipid nanoparticles (SLN) versus nanostructured lipid carriers (NLC). Eur J Pharm Biopharm 70(2):633–640

Fang R et al (2011) Bovine serum albumin nanoparticle promotes the stability of quercetin in simulated intestinal fluid. J Agric Food Chem 59(11):6292–6298

Fang B et al (2014) Bovine lactoferrin binds oleic acid to form an anti-tumor complex similar to HAMLET. Biochim Biophys Acta Mol Cell Biol Lipids 1841(4):535–543

Farjami T, Madadlou A, Labbafi M (2016) Modulating the textural characteristics of whey protein nanofibril gels with different concentrations of calcium chloride. J Dairy Res 83(1):109–114

Feng J-L et al (2015) Fabrication and characterization of stable soy β-conglycinin–dextran core–shell nanogels prepared via a self-assembly approach at the isoelectric point. J Agric Food Chem 63(26):6075–6083

Feng J et al (2016) Improved bioavailability of curcumin in ovalbumin-dextran nanogels prepared by Maillard reaction. J Funct Foods 27:55–68

Freitas RA (2005) Nanotechnology, nanomedicine and nanosurgery. Int J Surg 4(3):243–246

Froiio F et al (2019) Edible polymers for essential oils encapsulation: application in food preservation. Ind Eng Chem Res 58(46):20932–20945

Fuciños C et al (2017) Creating functional nanostructures: encapsulation of caffeine into α-lactalbumin nanotubes. Innovative Food Sci Emerg Technol 40:10–17

Fuciños C et al (2021) Biofunctionality assessment of α-lactalbumin nanotubes. Food Hydrocoll 117:106665

Gaba B et al (2015) Nanostructured lipid carrier system for topical delivery of terbinafine hydrochloride. Bull Facul Pharm Cairo Univ 53(2):147–159

Gagliardi A et al (2021) Zein-vs PLGA-based nanoparticles containing rutin: a comparative investigation. Mater Sci Eng C 118:111538

Gajic S, Ivana M et al (2021) Ultrasound-assisted extraction of carotenoids from orange peel using olive oil and its encapsulation in ca-alginate beads. Biomol Ther 11(2):225

Ganta S et al (2008) A review of stimuli-responsive nanocarriers for drug and gene delivery. J Control Release 126(3):187–204

Gaucheron F (2000) Iron fortification in dairy industry. Trends Food Sci Technol 11(11):403–409

Gelperina S et al (2005) The potential advantages of nanoparticle drug delivery systems in chemotherapy of tuberculosis. Am J Respir Crit Care Med 172(12):1487–1490

Geng XL et al (2016) Formation of nanotubes and gels at a broad pH range upon partial hydrolysis of bovine α-lactalbumin. Int Dairy J 52:72–81

Gesinde FA, Udechukwu MC, Aluko RE (2018) Structural and functional characterization of legume seed ferritin concentrates. J Food Biochem 42(3):e12498

Gharibzahedi SMT, Jafari SM (2017) The importance of minerals in human nutrition: bioavailability, food fortification, processing effects and nanoencapsulation. Trends Food Sci Technol 62:119–132

Ghasemi S et al (2018) Nanoencapsulation of d-limonene within nanocarriers produced by pectin-whey protein complexes. Food Hydrocoll 77:152–162

Ghayour N et al (2019) Nanoencapsulation of quercetin and curcumin in casein-based delivery systems. Food Hydrocoll 87:394–403

Ghorani B, Tucker N (2015) Fundamentals of electrospinning as a novel delivery vehicle for bioactive compounds in food nanotechnology. Food Hydrocoll 51:227–240

Ghosal K, Sarkar K (2018) Biomedical applications of graphene nanomaterials and beyond. ACS Biomater Sci Eng 4(8):2653–2703

Gianello R et al (2005) α-Tocopheryl phosphate: a novel, natural form of vitamin E. Free Radic Biol Med 39(7):970–976

Giessen TW, Silver PA (2017) Widespread distribution of encapsulin nanocompartments reveals functional diversity. Nat Microbiol 2(6):1–11

Glasser WG (2008) Cellulose and associated heteropolysaccharides. Glycoscience:1473. https://doi.org/10.1007/978-3-540-30429-6_36

Görner T et al (1999) Lidocaine-loaded biodegradable nanospheres. I. Optimization of the drug incorporation into the polymer matrix. J Control Release 57(3):259–268

Graveland-Bikker JF et al (2006) Growth and structure of α-lactalbumin nanotubes. J Appl Crystallogr 39(2):180–184

Graveland-Bikker JF et al (2009) Structural characterization of α-lactalbumin nanotubes. Soft Matter 5(10):2020–2026

Grumezescu AM (2017) Design of nanostructures for theranostics applications. William Andrew

Guo C et al (2010) Lyotropic liquid crystal systems in drug delivery. Drug Discov Today 15(23–24):1032–1040

Gupta A et al (2016a) Ultrastable and biofunctionalizable gold nanoparticles. ACS Appl Mater Interfaces 8(22):14096–14101

Gupta S et al (2016b) Encapsulation: entrapping essential oil/flavors/aromas in food. In: Encapsulations. Elsevier, pp 229–268

Gurunathan S et al (2018) Cytotoxicity and transcriptomic analysis of silver nanoparticles in mouse embryonic fibroblast cells. Int J Mol Sci 19(11):3618

Gurunathan S et al (2019) Evaluation of graphene oxide induced cellular toxicity and transcriptome analysis in human embryonic kidney cells. Nano 9(7):969

Hadiya S et al (2021) Nanoparticles integrating natural and synthetic polymers for in vivo insulin delivery. Pharm Dev Technol 26(1):30–40

Hao T et al (2020) Preparation, characterization, antioxidant evaluation of new curcumin derivatives and effects of forming HSA-bound nanoparticles on the stability and activity. Eur J Med Chem 207:112798

Harisa GI, Alomrani AH, Badran MM (2017) Simvastatin-loaded nanostructured lipid carriers attenuate the atherogenic risk of erythrocytes in hyperlipidemic rats. Eur J Pharm Sci 96:62–71

He X, Hwang H-M (2016) Nanotechnology in food science: functionality, applicability, and safety assessment. J Food Drug Anal 24(4):671–681

Hitchcock KE et al (2009) Delivery of targeted echogenic liposomes in an ex vivo mouse aorta model. J Acoust Soc Am 125(4):2713–2713

Horton S, Ross J (2003) The economics of iron deficiency. Food Policy 28(1):51–75

Hosseini SM et al (2013) Incorporation of essential oil in alginate microparticles by multiple emulsion/ionic gelation process. Int J Biol Macromol 62:582–588

Hosseini SF, Nahvi Z, Zandi M (2019) Antioxidant peptide-loaded electrospun chitosan/poly (vinyl alcohol) nanofibrous mat intended for food biopackaging purposes. Food Hydrocoll 89:637–648

Hu RL et al (2014) Inhibition effect of silver nanoparticles on herpes simplex virus 2. Genet Mol Res 13(3):7022–7028

Huang S-L et al (2001) Improving ultrasound reflectivity and stability of echogenic liposomal dispersions for use as targeted ultrasound contrast agents. J Pharm Sci 90(12):1917–1926

Huang S-L et al (2002) Physical correlates of the ultrasonic reflectivity of lipid dispersions suitable as diagnostic contrast agents. Ultrasound Med Biol 28(3):339–348

Huang W et al (2013) Drug-loaded zein nanofibers prepared using a modified coaxial electrospinning process. AAPS PharmSciTech 14(2):675–681

Huang H et al (2020) Ca 2+ participating self-assembly of an apoferritin nanostructure for nucleic acid drug delivery. Nanoscale 12(13):7347–7357

Huppertz T, de Kruif CG (2008) Structure and stability of nanogel particles prepared by internal cross-linking of casein micelles. Int Dairy J 18(5):556–565

Iacovacci V et al (2015) Untethered magnetic millirobot for targeted drug delivery. Biomed Microdevices 17(3):1–12

Iglič A, Rappolt M (2019) Advances in biomembranes and lipid self-assembly. Academic

Ilari A et al (1999) Crystallization and preliminary X-ray crystallographic analysis of the unusual ferritin from listeria innocua. Acta Crystallogr D Biol Crystallogr 55(2):552–553

Inal A, Yenipazar H, Şahin-Yeşilçubuk N (2022) Preparation and characterization of nanoemulsions of curcumin and echium oil. Heliyon 8(2):e08974

Jafari SM (2017) Nanoencapsulation of food bioactive ingredients: principles and applications. Academic

Jafari SM et al (2008) Re-coalescence of emulsion droplets during high-energy emulsification. Food Hydrocoll 22(7):1191–1202

Jafarifar Z et al (2022) Preparation and characterization of nanostructured lipid carrier (NLC) and nanoemulsion containing vitamin D3. Appl Biochem Biotechnol 194(2):914–929

Jahanshahi M, Babaei Z (2008) Protein nanoparticle: a unique system as drug delivery vehicles. Afr J Biotechnol 7(25). https://doi.org/10.4314/ajb.v7i25.59701

Jain P et al (2017) Nanostructure lipid carriers: a modish contrivance to overcome the ultraviolet effects. Egypt J Basic Appl Sci 4(2):89–100

Jain A et al (2018) Lycopene loaded whey protein isolate nanoparticles: an innovative endeavor for enhanced bioavailability of lycopene and anti-cancer activity. Int J Pharm 546(1–2):97–105

Jaiswal P, Gidwani B, Vyas A (2016) Nanostructured lipid carriers and their current application in targeted drug delivery. Artif Cells Nanomed Biotechnol 44:27–40

Jenning V, Thünemann AF, Gohla SH (2000) Characterisation of a novel solid lipid nanoparticle carrier system based on binary mixtures of liquid and solid lipids. Int J Pharm 199(2):167–177

Jiang S et al (2019) Pea protein nanoemulsion and nanocomplex as carriers for protection of cholecalciferol (vitamin D3). Food Bioprocess Technol 12(6):1031–1040

Jiang B et al (2020) A natural drug entry channel in the ferritin nanocage. Nano Today 35:100948

Jin B et al (2016) Self-assembled modified soy protein/dextran nanogel induced by ultrasonication as a delivery vehicle for riboflavin. Molecules 21(3):282

Jones JA, Giessen TW (2021) Advances in encapsulin nanocompartment biology and engineering. Biotechnol Bioeng 118(1):491–505

Jores K, Mehnert W, Mäder K (2003) Physicochemical investigations on solid lipid nanoparticles and on oil-loaded solid lipid nanoparticles: a nuclear magnetic resonance and electron spin resonance study. Pharm Res 20(8):1274–1283

Jutz G et al (2015) Ferritin: a versatile building block for bionanotechnology. Chem Rev 115(4):1653–1701

Kaddah S et al (2018) Cholesterol modulates the liposome membrane fluidity and permeability for a hydrophilic molecule. Food Chem Toxicol 113:40–48

Kapustová M et al (2021) Nanoencapsulated essential oils with enhanced antifungal activity for potential application on agri-food, material and environmental fields. Antibiotics 10(1):31

Karami Z, Hamidi M (2016) Cubosomes: remarkable drug delivery potential. Drug Discov Today 21(5):789–801

Karim M, Fathi M, Soleimanian-Zad S (2021) Nanoencapsulation of cinnamic aldehyde using zein nanofibers by novel needle-less electrospinning: production, characterization and their application to reduce nitrite in sausages. J Food Eng 288:110140

Katouzian I, Jafari SM (2016) Nano-encapsulation as a promising approach for targeted delivery and controlled release of vitamins. Trends Food Sci Technol 53:34–48

Katouzian I, Jafari SM (2019) Protein nanotubes as state-of-the-art nanocarriers: synthesis methods, simulation and applications. J Control Release 303:302–318

Keykhosravy K et al (2022) Protective effect of chitosan-loaded nanoemulsion containing Zataria multiflora Boiss and Bunium persicum Boiss essential oils as coating on lipid and protein oxidation in chill stored turkey breast fillets. J Food Sci 87(1):251–265

Kim KK, Kim R, Kim S-H (1998) Crystal structure of a small heat-shock protein. Nature 394(6693):595–599

Kim S et al (2010) β-CD-mediated encapsulation enhanced stability and solubility of astaxanthin. J Korean Soc Appl Biol Chem 53(5):559–565

Kiss É (2020) Nanotechnology in food systems: a review. Acta Aliment 49(4):460–474

Kraskiewicz H et al (2013) Assembly of protein-based hollow spheres encapsulating a therapeutic factor. ACS Chem Neurosci 4(9):1297–1304

Krishnamachari Y et al (2011) Nanoparticle delivery systems in cancer vaccines. Pharm Res 28(2):215–236

Kroes-Nijboer A, Venema P, van der Linden E (2012) Fibrillar structures in food. Food Funct 3(3):221–227

Kumar SS, Chauhan AS, Giridhar P (2020) Nanoliposomal encapsulation mediated enhancement of betalain stability: characterisation, storage stability and antioxidant activity of Basella rubra L. fruits for its applications in vegan gummy candies. Food Chem 333:127442

Kwekkeboom RFJ et al (2016) Increased local delivery of antagomir therapeutics to the rodent myocardium using ultrasound and microbubbles. J Control Release 222:18–31

Lassalle V, Ferreira ML (2007) PLA nano-and microparticles for drug delivery: an overview of the methods of preparation. Macromol Biosci 7(6):767–783

Lawson DM et al (1991) Solving the structure of human H ferritin by genetically engineering intermolecular crystal contacts. Nature 349(6309):541–544

Lee CC et al (2005) Designing dendrimers for biological applications. Nat Biotechnol 23(12):1517–1526

Leonetti J-P et al (1990) Antibody-targeted liposomes containing oligodeoxyribonucleotides complementary to viral RNA selectively inhibit viral replication. Proc Natl Acad Sci 87(7):2448–2451

Li W et al (2009) Preparation and characterization of gelatin/SDS/NaCMC microcapsules with compact wall structure by complex coacervation. Colloids Surf A Physicochem Eng Asp 333(1–3):133–137

Li Z et al (2015) Self-assembled lysozyme/carboxymethylcellulose nanogels for delivery of methotrexate. Int J Biol Macromol 75:166–172

Li H et al (2016) Electrospun gelatin nanofibers loaded with vitamins A and E as antibacterial wound dressing materials. RSC Adv 6(55):50267–50277

Li D et al (2019a) The characterization and stability of the soy protein isolate/1-Octacosanol nanocomplex. Food Chem 297:124766

Li X et al (2019b) Development of hollow kafirin-based nanoparticles fabricated through layer-by-layer assembly as delivery vehicles for curcumin. Food Hydrocoll 96:93–101

Li H et al (2020) Purification and characterizations of a nanocage ferritin GF1 from oyster (Crassostrea gigas). LWT 127:109416

Li J et al (2021a) Surface properties and liquid crystal properties of Alkyltetra (oxyethyl) β-d-Glucopyranoside. J Agric Food Chem 69(36):10617–10629

Li M et al (2021b) Fabrication of eugenol loaded gelatin nanofibers by electrospinning technique as active packaging material. LWT 139:110800

Li Z et al (2022) Chitosan/zein films incorporated with essential oil nanoparticles and nanoemulsions: similarities and differences. Int J Biol Macromol 208:983–994

Libinaki R et al (2010) Effect of tocopheryl phosphate on key biomarkers of inflammation: implication in the reduction of atherosclerosis progression in a hypercholesterolaemic rabbit model. Clin Exp Pharmacol Physiol 37(5–6):587–592

Lin L et al (2015) Construction of pH-sensitive lysozyme/pectin nanogel for tumor methotrexate delivery. Colloids Surf B: Biointerfaces 126:459–466

Liu Z et al (2008) Polysaccharides-based nanoparticles as drug delivery systems. Adv Drug Deliv Rev 60(15):1650–1662

Liu Y et al (2017) Improved antioxidant activity and physicochemical properties of curcumin by adding ovalbumin and its structural characterization. Food Hydrocoll 72:304–311

Liu Y et al (2018a) Enhanced pH and thermal stability, solubility and antioxidant activity of resveratrol by nanocomplexation with α-lactalbumin. Food Funct 9(9):4781–4790

Liu Y et al (2018b) Ovalbumin as a carrier to significantly enhance the aqueous solubility and photostability of curcumin: interaction and binding mechanism study. Int J Biol Macromol 116:893–900

Liu T et al (2021) Fabrication and comparison of active films from chitosan incorporating different spice extracts for shelf life extension of refrigerated pork. LWT 135:110181

Loo CH et al (2013) Effect of compositions in nanostructured lipid carriers (NLC) on skin hydration and occlusion. Int J Nanomedicine 8:13

López-García R, Ganem-Rondero A (2015) Solid lipid nanoparticles (SLN) and nanostructured lipid carriers (NLC): occlusive effect and penetration enhancement ability. JCDSA 5(02):62–72

Lopez-Garcia E et al (2005) Consumption of trans fatty acids is related to plasma biomarkers of inflammation and endothelial dysfunction. J Nutr 135(3):562–566

Lou XW, Archer LA, Yang Z (2008) Hollow micro−/nanostructures: synthesis and applications. Adv Mater 20(21):3987–4019

Lu GW, Gao P (2010) Emulsions and microemulsions for topical and transdermal drug delivery. In: Handbook of non-invasive drug delivery systems. Elsevier, pp 59–94

Luo R et al (2020) Genipin-crosslinked human serum albumin coating using a tannic acid layer for enhanced oral administration of curcumin in the treatment of ulcerative colitis. Food Chem 330:127241

Lv C et al (2021) Redesign of protein nanocages: the way from 0D, 1D, 2D to 3D assembly. Chem Soc Rev 50(6):3957–3989

Ma X et al (2021) Ethyl cellulose particles loaded with α-tocopherol for inhibiting thermal oxidation of soybean oil. Carbohydr Polym 252:117169

Madadlou A, Jaberipour S, Eskandari MH (2014) Nanoparticulation of enzymatically cross-linked whey proteins to encapsulate caffeine via microemulsification/heat gelation procedure. LWT-Food Sci Technol 57(2):725–730

Maestrelli F et al (2006) A new drug nanocarrier consisting of chitosan and hydoxypropylcyclodextrin. Eur J Pharm Biopharm 63(2):79–86

Mahapatro A, Singh DK (2011) Biodegradable nanoparticles are excellent vehicle for site directed in-vivo delivery of drugs and vaccines. J Nanobiotechnol 9:1–11

Maherani B et al (2012) Influence of lipid composition on physicochemical properties of nanoliposomes encapsulating natural dipeptide antioxidant l-carnosine. Food Chem 134(2):632–640

Maldonado L, Kokini J (2018) An optimal window for the fabrication of edible polyelectrolyte complex nanotubes (EPCNs) from bovine serum albumin (BSA) and sodium alginate. Food Hydrocoll 77:336–346

Maleki G, Woltering EJ, Mozafari MR (2022) Applications of chitosan-based carrier as an encapsulating agent in food industry. Trends Food Sci Technol 120:88–99

Martin FJ, Heath TD, New RRC (1990) Covalent attachment of proteins to liposomes. In Liposomes: a practical approach, p 1

Matencio A et al (2021) Nanoparticles of betalamic acid derivatives with cyclodextrins. Physicochemistry, production characterization and stability. Food Hydrocoll 110:106176

Mattevi A et al (1992) Atomic structure of the cubic core of the pyruvate dehydrogenase multienzyme Eomplex. Science 255(5051):1544–1550

McClements DJ (2004) Food emulsions: principles, practices, and techniques. CRC Press

McClements DJ (2012) Nanoemulsions versus microemulsions: terminology, differences, and similarities. Soft Matter 8(6):1719–1729

McClements DJ (2015) Reduced-fat foods: the complex science of developing diet-based strategies for tackling overweight and obesity. Adv Nutr 6(3):338S–352S

McClements DJ et al (2009) Structural design principles for delivery of bioactive components in nutraceuticals and functional foods. Crit Rev Food Sci Nutr 49(6):577–606

McHugh CA et al (2014) A virus capsid-like nanocompartment that stores iron and protects bacteria from oxidative stress. EMBO J 33(17):1896–1911

Md S et al (2019) Nanoencapsulation of betamethasone valerate using high pressure homogenization–solvent evaporation technique: optimization of formulation and process parameters for efficient dermal targeting. Drug Dev Ind Pharm 45(2):323–332

Mehrad B et al (2015) Characterization of dried fish oil from menhaden encapsulated by spray drying. Aquac Aquar Conserv Legis 8(1):57–69

Mehrnia M-A et al (2016) Crocin loaded nano-emulsions: factors affecting emulsion properties in spontaneous emulsification. Int J Biol Macromol 84:261–267

Mehrnia M-A et al (2017) Rheological and release properties of double nano-emulsions containing crocin prepared with Angum gum, Arabic gum and whey protein. Food Hydrocoll 66:259–267

Mendes AC, Stephansen K, Chronakis IS (2017) Electrospinning of food proteins and polysaccharides. Food Hydrocoll 68:53–68

Meng D et al (2019) Influence of manothermosonication on the physicochemical and functional properties of ferritin as a nanocarrier of iron or bioactive compounds. J Agric Food Chem 67(23):6633–6641

Meng R et al (2021) Preparation and characterization of zein/carboxymethyl dextrin nanoparticles to encapsulate curcumin: physicochemical stability, antioxidant activity and controlled release properties. Food Chem 340:127893

Merkel TJ et al (2011) Using mechanobiological mimicry of red blood cells to extend circulation times of hydrogel microparticles. Proc Natl Acad Sci 108(2):586–591

Mertins O, Mathews PD, Angelova A (2020) Advances in the design of ph-sensitive cubosome liquid crystalline nanocarriers for drug delivery applications. Nano 10(5):963

Michalet X et al (2005) Quantum dots for live cells, in vivo imaging, and diagnostics. Science 307(5709):538–544

Milne JLS et al (2006) Molecular structure of a 9-MDa icosahedral pyruvate dehydrogenase subcomplex containing the E2 and E3 enzymes using cryoelectron microscopy. J Biol Chem 281(7):4364–4370

Min T et al (2021) Novel antimicrobial packaging film based on porous poly (lactic acid) nanofiber and polymeric coating for humidity-controlled release of thyme essential oil. LWT 135:110034

Minato T et al (2020) Biochemical and structural characterization of a thermostable Dps protein with his-type ferroxidase centers and outer metal-binding sites. FEBS Open Bio 10(7):1219–1229

Mo J, Milleret G, Nagaraj M (2017) Liquid crystal nanoparticles for commercial drug delivery. Liq Cryst Rev 5(2):69–85

Mohammadi A et al (2016a) Nano-encapsulation of olive leaf phenolic compounds through WPC–pectin complexes and evaluating their release rate. Int J Biol Macromol 82:816–822

Mohammadi A et al (2016b) Application of nano-encapsulated olive leaf extract in controlling the oxidative stability of soybean oil. Food Chem 190:513–519

Mohammadian M, Madadlou A (2016) Characterization of fibrillated antioxidant whey protein hydrolysate and comparison with fibrillated protein solution. Food Hydrocoll 52:221–230

Mohammadian M, Madadlou A (2018) Technological functionality and biological properties of food protein nanofibrils formed by heating at acidic condition. Trends Food Sci Technol 75:115–128

Mohammadian M et al (2019) Enhancing the aqueous solubility of curcumin at acidic condition through the complexation with whey protein nanofibrils. Food Hydrocoll 87:902–914

Mohammadian M et al (2020) Nanostructured food proteins as efficient systems for the encapsulation of bioactive compounds. Food Sci Human Wellness 9(3):199–213

Montanari D et al (2015) Development of membrane cryostats for large liquid argon neutrino detectors. IOP Conf Ser Mater Sci Eng 101:012049

Montemiglio LC et al (2019) Cryo-EM structure of the human ferritin–transferrin receptor 1 complex. Nat Commun 10(1):1–8

Mozafari RM (2005) Nanoliposomes: from fundamentals to recent developments. Trafford

Mozafari MR et al (2008) Encapsulation of food ingredients using nanoliposome technology. Int J Food Prop 11(4):833–844

Mozafari MR, Javanmard R, Raji M (2017) Tocosome: novel drug delivery system containing phospholipids and tocopheryl phosphates. Int J Pharm 528(1–2):381–382

Müller RH, Keck CM (2015) Next generation after SLN® and NLC®–the "chaotic" smartLipids®. Wissenschaftliche Posterausstellung:9

Muller RH, Runge SA (1998) Solid lipid nanoparticles (SLN) for controlled drug delivery. In: Submicron emulsions in drug targeting and delivery, vol 219, p 234

Müller RH, Ruick R, Keck CM (2014a) smartLipids®—the new generation of lipid nanoparticles after SLN and NLC. In: Proceedings of the annual meeting of American association of pharmaceutical scientists, pp 2–6

Müller RH, Ruick R, Keck CM (2014b) smartLipids®—the next generation of lipid nanoparticles by optimized design of particle matrix. In: Proceedings of the annual meeting of German pharmaceutical society, pp 24–26

Munteanu A et al (2004) Modulation of cell proliferation and gene expression by α-tocopheryl phosphates: relevance to atherosclerosis and inflammation. Biochem Biophys Res Commun 318(1):311–316

Murgia S, Biffi S, Mezzenga R (2020) Recent advances of non-lamellar lyotropic liquid crystalline nanoparticles in nanomedicine. Curr Opin Colloid Interface Sci 48:28–39

Naranjo-Durán AM et al (2021) Modified-release of encapsulated bioactive compounds from annatto seeds produced by optimized ionic gelation techniques. Sci Rep 11(1):1–10

Nasrollahi F et al (2020) Incorporation of graphene quantum dots, iron, and doxorubicin in/on ferritin nanocages for bimodal imaging and drug delivery. Adv Ther 3(3):1900183

Nastruzzi C (2004) Lipospheres in drug targets and delivery: approaches, methods, and applications. CRC Press

Nasu E, Kawakami N, Miyamoto K (2021) Nanopore-controlled dual-surface modifications on artificial protein nanocages as nanocarriers. ACS Appl Nano Mater 4(3):2434–2439

Neto AMJC, Aragao Lopes I, Pirota KR (2010) A review on nanorobotics. J Comput Theor Nanosci 7(10):1870–1877

Nishio K et al (2011) α-Tocopheryl phosphate: uptake, hydrolysis, and antioxidant action in cultured cells and mouse. Free Radic Biol Med 50(12):1794–1800

Nowicka A et al (2019) Comparison of polyphenol content and antioxidant capacity of strawberry fruit from 90 cultivars of Fragaria× ananassa Duch. Food Chem 270:32–46

Ogru E et al (2003) Vitamin E phosphate: an endogenous form of vitamin E. Medimond Srl 2003:127–132

Ojagh SM, Hasani S (2018) Characteristics and oxidative stability of fish oil nano-liposomes and its application in functional bread. J Food Meas Charact 12(2):1084–1092

Palla CA et al (2022) Preparation of highly stable oleogel-based nanoemulsions for encapsulation and controlled release of curcumin. Food Chem 378:132132

Pan J et al (2019) Development of polyvinyl alcohol/β-cyclodextrin antimicrobial nanofibers for fresh mushroom packaging. Food Chem 300:125249

Panyam J, Labhasetwar V (2003) Biodegradable nanoparticles for drug and gene delivery to cells and tissue. Adv Drug Deliv Rev 55(3):329–347

Pardeike J, Hommoss A, Müller RH (2009) Lipid nanoparticles (SLN, NLC) in cosmetic and pharmaceutical dermal products. Int J Pharm 366(1–2):170–184

Park H, Park K (1996) Biocompatibility issues of implantable drug delivery systems. Pharm Res 13(12):1770–1776

Patel GM et al (2006) Nanorobot: a versatile tool in nanomedicine. J Drug Target 14(2):63–67

Patel D et al (2012) Nanostructured lipid carriers (NLC)-based gel for the topical delivery of aceclofenac: preparation, characterization, and in vivo evaluation. Sci Pharm 80(3):749–764

Pathak L, Kanwal A, Agrawal Y (2015) Curcumin loaded self assembled lipid-biopolymer nanoparticles for functional food applications. J Food Sci Technol 52(10):6143–6156

Patil P, Killedar S (2021) Green approach towards synthesis and characterization of GMO/chitosan nanoparticles for in vitro release of quercetin: isolated from peels of pomegranate fruit. J Pharm Innov 17:1–14

Perinelli DR et al (2020) Encapsulation of flavours and fragrances into polymeric capsules and cyclodextrins inclusion complexes: an update. Molecules 25(24):5878

Peschka R, Dennehy C, Szoka Jr FC (1998) A simple in vitro model to study the release kinetics of liposome encapsulated material. J Control Release 56(1–3):41–51

Peters RJB et al (2016) Nanomaterials for products and application in agriculture, feed and food. Trends Food Sci Technol 54:155–164

Piskula MK, Murota K, Terao J (2012) Bioavailability of flavonols and flavones. Flavonoids and related compounds: bioavailability and function. CRC Press, Boca Raton, pp 93–107

Plourde M, Cunnane SC (2007) Extremely limited synthesis of long chain polyunsaturates in adults: implications for their dietary essentiality and use as supplements. Appl Physiol Nutr Metab 32(4):619–634

Prasad Reddy MN, Padma Ishwarya S, Anandharamakrishnan C (2019) Nanoencapsulation of roasted coffee bean oil in whey protein wall system through nanospray drying. J Food Process Pres 43(3):e13893

Prokop A, Davidson JM (2008) Nanovehicular intracellular delivery systems. J Pharm Sci 97(9):3518–3590

Pujara N et al (2017) Enhanced colloidal stability, solubility and rapid dissolution of resveratrol by nanocomplexation with soy protein isolate. J Colloid Interface Sci 488:303–308

Pyo S-M, Müller RH, Keck CM (2017) Encapsulation by nanostructured lipid carriers. In: Nanoencapsulation technologies for the food and nutraceutical industries. Elsevier, pp 114–137

Qiu H, Caffrey M (2000) The phase diagram of the monoolein/water system: metastability and equilibrium aspects. Biomaterials 21(3):223–234

Rahul VA (2017) A brief review on nanorobots. SSRG-IJME 4:15–21

Rajinikanth PS, Chellian J (2016) Development and evaluation of nanostructured lipid carrier-based hydrogel for topical delivery of 5-fluorouracil. Int J Nanomedicine 11:5067–5077

Ramos PE et al (2016) Development of an immobilization system for in situ micronutrients release. Food Res Int 90:121–132

Ramos OL et al (2017) Design of whey protein nanostructures for incorporation and release of nutraceutical compounds in food. Crit Rev Food Sci Nutr 57(7):1377–1393

Rao J, McClements DJ (2011) Food-grade microemulsions, nanoemulsions and emulsions: fabrication from sucrose monopalmitate & lemon oil. Food Hydrocoll 25(6):1413–1423

Rashid R et al (2022) Ultrasound assisted extraction of bioactive compounds from pomegranate peel, their nanoencapsulation and application for improvement in shelf life extension of edible oils. Food Chem 385:132608

Rashidi L, Khosravi-Darani K (2011) The applications of nanotechnology in food industry. Crit Rev Food Sci Nutr 51(8):723–730

Rasti B et al (2012) Comparative study of the oxidative and physical stability of liposomal and nanoliposomal polyunsaturated fatty acids prepared with conventional and Mozafari methods. Food Chem 135(4):2761–2770

Reboredo C et al (2021) Zein-based nanoparticles as oral carriers for insulin delivery. Pharmaceutics 14(1):39

Rehan F, Ahemad N, Gupta M (2019) Casein nanomicelle as an emerging biomaterial—a comprehensive review. Colloids Surf B: Biointerfaces 179:280–292

Reis CP et al (2007) Nanoparticulate delivery system for insulin: design, characterization and in vitro/in vivo bioactivity. Eur J Pharm Sci 30(5):392–397

Ren D, Kratz F, Wang S-W (2011) Protein nanocapsules containing doxorubicin as a pH-responsive delivery system. Small 7(8):1051–1060

Ren D et al (2012) Biomimetic design of protein nanomaterials for hydrophobic molecular transport. Adv Funct Mater 22(15):3170–3180

Revathi S, Dhanaraju MD (2014) Hexosomes as a novel drug delivery system: a review. Int J Pharmacol Pharm Sci 6:58–63

Roohinejad S, Greiner R, Oey I, Wen J (eds) (2018) Emulsion-based systems for delivery of food active compounds: formation, application, health and safety. Wiley

Rosa RM et al (2017) Simultaneous photo-induced cross-linking and silver nanoparticle formation in a PVP electrospun wound dressing. Mater Lett 207:145–148

Rostamabadi H et al (2020) Electrospinning approach for nanoencapsulation of bioactive compounds; recent advances and innovations. Trends Food Sci Technol 100:190–209

Ruengdech A, Siripatrawan U (2021) Application of catechin nanoencapsulation with enhanced antioxidant activity in high pressure processed catechin-fortified coconut milk. LWT 140:110594

Sadeghi R et al (2013) Biocompatible nanotubes as potential carrier for curcumin as a model bioactive compound. J Nanopart Res 15(11):1–11

Sadeghi R et al (2014) The effect of different desolvating agents on BSA nanoparticle properties and encapsulation of curcumin. J Nanopart Res 16(9):1–14

Sagalowicz L, Leser ME (2010) Delivery systems for liquid food products. Curr Opin Colloid Interface Sci 15(1–2):61–72

Saha M (2009) Nanomedicine: promising tiny machine for the healthcare in future-a review. Oman Med J 24(4):242–247

Sahafi SM et al (2021) Pomegranate seed oil nanoemulsion enriched by α-tocopherol; the effect of environmental stresses and long-term storage on its physicochemical properties and oxidation stability. Food Chem 345:128759

Said M et al (2021) Central composite optimization of ocular mucoadhesive cubosomes for enhanced bioavailability and controlled delivery of voriconazole. J Drug Deliv Sci Technol 61:102075

Saitoh Y, Yumoto A, Miwa N (2009) α-Tocopheryl phosphate suppresses tumor invasion concurrently with dynamic morphological changes and delocalization of cortactin from invadopodia. Int J Oncol 35(6):1277–1288

Sáiz-Abajo M-J et al (2013) Thermal protection of β-carotene in re-assembled casein micelles during different processing technologies applied in food industry. Food Chem 138(2–3):1581–1587

Santana RC, Perrechil FA, Cunha RL (2013) High-and low-energy emulsifications for food applications: a focus on process parameters. Food Eng Rev 5(2):107–122

Schleeger M et al (2013) Amyloids: from molecular structure to mechanical properties. Polymer 54(10):2473–2488

Schmidt CK et al (2020) Engineering microrobots for targeted cancer therapies from a medical perspective. Nat Commun 11(1):5618

Selvamuthukumar S, Velmurugan R (2012) Nanostructured lipid carriers: a potential drug carrier for cancer chemotherapy. Lipids Health Dis 11(1):1–8

Semo E et al (2007) Casein micelle as a natural nano-capsular vehicle for nutraceuticals. Food Hydrocoll 21(5–6):936–942

Serfert Y et al (2014) Characterisation and use of β-lactoglobulin fibrils for microencapsulation of lipophilic ingredients and oxidative stability thereof. J Food Eng 143:53–61

Seyedabadi MM et al (2021) Development and characterization of chitosan-coated nanoliposomes for encapsulation of caffeine. Food Biosci 40:100857

Shah R et al (2015) Lipid nanoparticles: production, characterization and stability, vol 1. Springer

Shahidi F, Hossain A (2018) Bioactives in spices, and spice oleoresins: phytochemicals and their beneficial effects in food preservation and health promotion. J Food Bioact 3:8–75

Shi X et al (2017) Comparative studies on glycerol monooleate-and phytantriol-based cubosomes containing oridonin in vitro and in vivo. Pharm Dev Technol 22(3):322–329

Shimoni E (2009) Nanotechnology for foods: delivery systems. In: Global issues in food science and technology. Elsevier, pp 411–424

Sim S, Aida T (2017) Swallowing a surgeon: toward clinical nanorobots. Acc Chem Res 50(3):492–497

Simionato I et al (2019) Encapsulation of cinnamon oil in cyclodextrin nanosponges and their potential use for antimicrobial food packaging. Food Chem Toxicol 132:110647

Simiqueli AA et al (2019) The W/O/W emulsion containing FeSO4 in the different phases alters the hedonic thresholds in milk-based dessert. LWT 99:98–104

Singh R, Lillard Jr JW (2009) Nanoparticle-based targeted drug delivery. Exp Mol Pathol 86(3):215–223

Singh J, Kaur K, Kumar P (2018) Optimizing microencapsulation of α-tocopherol with pectin and sodium alginate. J Food Sci Technol 55(9):3625–3631

Smith DAB et al (2007) Echogenic liposomes loaded with recombinant tissue-type plasminogen activator (rt-PA) for image-guided, ultrasound-triggered drug release. J Acoust Soc Am 122(5):3007–3007

Soto F et al (2020) Medical micro/nanorobots in precision medicine. Adv Sci 7(21):2002203

Souza C et al (2014) Mucoadhesive system formed by liquid crystals for buccal administration of poly (hexamethylene biguanide) hydrochloride. J Pharm Sci 103(12):3914–3923

Stewart PL (2017) Cryo-electron microscopy and cryo-electron tomography of nanoparticles. Wiley Interdiscip Rev Nanomed Nanobiotechnol 9(2):e1417

Sullivan ST et al (2014) Electrospinning and heat treatment of whey protein nanofibers. Food Hydrocoll 35:36–50

Sun L et al (2015) Soy protein isolate/cellulose nanofiber complex gels as fat substitutes: rheological and textural properties and extent of cream imitation. Cellulose 22(4):2619–2627

Sun W et al (2016) Bone-targeted mesoporous silica nanocarrier anchored by zoledronate for cancer bone metastasis. Langmuir 32(36):9237–9244

Sutter M et al (2008) Structural basis of enzyme encapsulation into a bacterial nanocompartment. Nat Struct Mol Biol 15(9):939–947

Suzuki T et al (2021) DNA-binding protein from starvation cells traps intracellular free-divalent iron and plays an important role in oxidative stress resistance in Acetobacter pasteurianus NBRC 3283. J Biosci Bioeng 131(3):256–263

Tan C et al (2015) Biopolymer–lipid bilayer interaction modulates the physical properties of liposomes: mechanism and structure. J Agric Food Chem 63(32):7277–7285

Tan C et al (2016a) Biopolymer-coated liposomes by electrostatic adsorption of chitosan (chitosomes) as novel delivery systems for carotenoids. Food Hydrocoll 52:774–784

Tan C et al (2016b) Polysaccharide-based nanoparticles by chitosan and gum arabic polyelectrolyte complexation as carriers for curcumin. Food Hydrocoll 57:236–245

Tang C-h (2021) Strategies to utilize naturally occurring protein architectures as nanovehicles for hydrophobic nutraceuticals. Food Hydrocoll 112:106344

Tang Y et al (2019) Electrospun gelatin nanofibers encapsulated with peppermint and chamomile essential oils as potential edible packaging. J Agric Food Chem 67(8):2227–2234

Tapal A, Tiku PK (2012) Complexation of curcumin with soy protein isolate and its implications on solubility and stability of curcumin. Food Chem 130(4):960–965

Tarhan Ö, Harsa Ş (2014) Nanotubular structures developed from whey-based α-lactalbumin fractions for food applications. Biotechnol Prog 30(6):1301–1310

Tarhan Ö, Tarhan E, Harsa Ş (2014) Investigation of the structure of alpha-lactalbumin protein nanotubes using optical spectroscopy. J Dairy Res 81(1):98–106

Tarhini M, Greige-Gerges H, Elaissari A (2017) Protein-based nanoparticles: from preparation to encapsulation of active molecules. Int J Pharm 522(1–2):172–197

Tavassoli-Kafrani E, Goli SAH, Fathi M (2018) Encapsulation of orange essential oil using cross-linked electrospun gelatin nanofibers. Food Bioprocess Technol 11(2):427–434

Taylor TM et al (2007) Characterization of antimicrobial-bearing liposomes by ζ-potential, vesicle size, and encapsulation efficiency. Food Biophys 2(1):1–9

Teixé-Roig J et al (2018) The effect of sodium carboxymethylcellulose on the stability and bioaccessibility of anthocyanin water-in-oil-in-water emulsions. Food Bioprocess Technol 11(12):2229–2241

Tenchov R et al (2021) Lipid nanoparticles—from liposomes to mRNA vaccine delivery, a landscape of research diversity and advancement. ACS Nano 15(11):16982–17015

Teng Z, Luo Y, Wang Q (2012) Nanoparticles synthesized from soy protein: preparation, characterization, and application for nutraceutical encapsulation. J Agric Food Chem 60(10):2712–2720

Thakuria R et al (2013) Pharmaceutical cocrystals and poorly soluble drugs. Int J Pharm 453(1):101–125

Thangavel K et al (2014) A survey on nano-robotics in nano-medicine. Nanotechnology 8(9):525–528

Thatipamula RP et al (2011) Formulation and in vitro characterization of domperidone loaded solid lipid nanoparticles and nanostructured lipid carriers. Daru 19(1):23–32

Thompson RF et al (2016) An introduction to sample preparation and imaging by cryo-electron microscopy for structural biology. Methods 100:3–15

Toldrá F et al (2020) Bioactive peptides generated in the processing of dry-cured ham. Food Chem 321:126689

Torchilin VP (2001) Structure and design of polymeric surfactant-based drug delivery systems. J Control Release 73(2–3):137–172

Torchilin VP (2008) Cell penetrating peptide-modified pharmaceutical nanocarriers for intracellular drug and gene delivery. Pept Sci 90(5):604–610

Tran N et al (2018) Manipulating the ordered nanostructure of self-assembled monoolein and phy-tantriol nanoparticles with unsaturated fatty acids. Langmuir 34(8):2764–2773

Trifković K et al (2016) Novel approaches in nanoencapsulation of aromas and flavors. Encapsulations:363–419

Upadhyay VP et al (2017) Nano robots in medicine: a review. Int J Eng Technol Manag Res 4(12):27–37

Uprit S et al (2013) Preparation and characterization of minoxidil loaded nanostructured lipid carrier gel for effective treatment of alopecia. Saudi Pharm J 21(4):379–385

Vafania B, Fathi M, Soleimanian-Zad S (2019) Nanoencapsulation of thyme essential oil in chitosan-gelatin nanofibers by nozzle-less electrospinning and their application to reduce nitrite in sausages. Food Bioprod Process 116:240–248

Vahidmoghadam F et al (2019) Characteristics of freeze-dried nanoencapsulated fish oil with whey protein concentrate and gum arabic as wall materials. Food Sci Technol 39:475–481

van Nieuwenhuyzen W, Szuhaj BF (1998) Effects of lecithins and proteins on the stability of emulsions. Lipid/Fett 100(7):282–291

Vasuki Y et al (2014) Semi-automatic mapping of geological structures using UAV-based photogrammetric data: an image analysis approach. Comput Geosci 69:22–32

Vickers NJ (2017) Animal communication: when i'm calling you, will you answer too? Curr Biol 27(14):R713–R715

Vignoli JA et al (2014) Roasting process affects differently the bioactive compounds and the antioxidant activity of arabica and robusta coffees. Food Res Int 61:279–285

Visentini FF et al (2017) Formation and colloidal stability of ovalbumin-retinol nanocomplexes. Food Hydrocoll 67:130–138

Wang W, Zhou C (2021) A journey of nanomotors for targeted cancer therapy: principles, challenges, and a critical review of the state-of-the-art. Adv Healthc Mater 10(2):2001236

Wang K et al (2016a) Self-assembled IR780-loaded transferrin nanoparticles as an imaging, targeting and PDT/PTT agent for cancer therapy. Sci Rep 6(1):1–11

Wang X et al (2016b) Synthesis, properties, and applications of hollow micro−/nanostructures. Chem Rev 116(18):10983–11060

Wang Y et al (2016c) Nanogels fabricated from bovine serum albumin and chitosan via self-assembly for delivery of anticancer drug. Colloids Surf B: Biointerfaces 146:107–113

Wang W et al (2019) AB loop engineered ferritin nanocages for drug loading under benign experimental conditions. Chem Commun 55(82):12344–12347

Wang S et al (2021) Nanocomplexes derived from chitosan and whey protein isolate enhance the thermal stability and slow the release of anthocyanins in simulated digestion and prepared instant coffee. Food Chem 336:127707

Wei Y et al (2018) β-Lactoglobulin as a nanotransporter for glabridin: exploring the binding properties and bioactivity influences. ACS Omega 3(9):12246–12252

Wen P et al (2017) Electrospinning: a novel nano-encapsulation approach for bioactive compounds. Trends Food Sci Technol 70:56–68

Wong AD, DeWit MA, Gillies ER (2012) Amplified release through the stimulus triggered degradation of self-immolative oligomers, dendrimers, and linear polymers. Adv Drug Deliv Rev 64(11):1031–1045

Wongsasulak S et al (2010) Electrospinning of food-grade nanofibers from cellulose acetate and egg albumen blends. J Food Eng 98(3):370–376

Wu D, Wan M (2008) A novel fluoride anion modified gelatin nanogel system. J Pharm Pharm Sci 11(4):32–45

Xavier LO et al (2021) Chitosan packaging functionalized with Cinnamodendron dinisii essential oil loaded zein: a proposal for meat conservation. Int J Biol Macromol 169:183–193

Xia S, Shiying X (2005) Ferrous sulfate liposomes: preparation, stability and application in fluid milk. Food Res Int 38(3):289–296

Xiao N-Y et al (2020) Construction of EVA/chitosan based PEG-PCL micelles nanocomposite films with controlled release of iprodione and its application in pre-harvest treatment of grapes. Food Chem 331:127277

Xu H et al (2011) Hollow nanoparticles from zein for potential medical applications. J Mater Chem 21(45):18227–18235

Xu H et al (2013) Biodegradable hollow zein nanoparticles for removal of reactive dyes from wastewater. J Environ Manag 125:33–40

Xu H et al (2015) Controlled delivery of hollow corn protein nanoparticles via non-toxic crosslinking: in vivo and drug loading study. Biomed Microdevices 17(1):1–8

Yaghmur A (2019) Nanoencapsulation of food ingredients by cubosomes and hexosomes. In: Lipid-based nanostructures for food encapsulation purposes. Elsevier, pp 483–522

Yaghmur A, Huiling M (2021) Recent advances in drug delivery applications of cubosomes, hexosomes, and solid lipid nanoparticles. Acta Pharm Sin B 11(4):871–885

Yan B et al (2021) Improvement of vitamin C stability in vitamin gummies by encapsulation in casein gel. Food Hydrocoll 113:106414

Yang R et al (2015) Ferritin, a novel vehicle for iron supplementation and food nutritional factors encapsulation. Trends Food Sci Technol 44(2):189–200

Yoksan R, Jirawutthiwongchai J, Arpo K (2010) Encapsulation of ascorbyl palmitate in chitosan nanoparticles by oil-in-water emulsion and ionic gelation processes. Colloids Surf B: Biointerfaces 76(1):292–297

Yu Z et al (2014) Bovine serum albumin nanoparticles as controlled release carrier for local drug delivery to the inner ear. Nanoscale Res Lett 9(1):1–7

Yuan L et al (2022) Making of Massoia lactone-loaded and food-grade nanoemulsions and their bioactivities against a pathogenic yeast. J Mar Sci Eng 10(3):339

Yurdugul S, Mozafari MR (2004) Recent advances in micro-and nanoencapsulation of food ingredients. Cell Mol Biol Lett 9(S2):64–65

Zabara A, Mezzenga R (2014) Controlling molecular transport and sustained drug release in lipid-based liquid crystalline mesophases. J Control Release 188:31–43

Zanetti M et al (2019) Encapsulation of geranyl cinnamate in polycaprolactone nanoparticles. Mater Sci Eng C 97:198–207

Zarrabi A et al (2020) Nanoliposomes and tocosomes as multifunctional nanocarriers for the encapsulation of nutraceutical and dietary molecules. Molecules 25(3):638

Zhai J et al (2019) Non-lamellar lyotropic liquid crystalline lipid nanoparticles for the next generation of nanomedicine. ACS Nano 13(6):6178–6206

Zhan F et al (2020) Complexation between sodium caseinate and gallic acid: effects on foam properties and interfacial properties of foam. Food Hydrocoll 99:105365

Zhang D, Dougherty SA, Liang J (2011) Fabrication of bovine serum albumin nanotubes through template-assisted layer by layer assembly. J Nanopart Res 13(4):1563–1571

Zhang T et al (2014) Encapsulation of anthocyanin molecules within a ferritin nanocage increases their stability and cell uptake efficiency. Food Res Int 62:183–192

Zhang H et al (2016a) New progress and prospects: the application of nanogel in drug delivery. Mater Sci Eng C 60:560–568

Zhang S et al (2016b) Conversion of the native 24-mer ferritin Nanocage into its non-native 16-mer analogue by insertion of extra amino acid residues. Angew Chem Int Ed 55(52):16064–16070

Zhang X-F et al (2016c) Silver nanoparticles: synthesis, characterization, properties, applications, and therapeutic approaches. Int J Mol Sci 17(9):1534

Zhang H, Li X, Kang H (2019a) Chitosan coatings incorporated with free or nano-encapsulated paulownia Tomentosa essential oil to improve shelf-life of ready-to-cook pork chops. LWT 116:108580

Zhang X et al (2019b) Thermostability of protein nanocages: the effect of natural extra peptide on the exterior surface. RSC Adv 9(43):24777–24782

Zhang L et al (2020) Forming nanoconjugates or inducing macroaggregates, curcumin dose effect on myosin assembling revealed by molecular dynamics simulation. Colloids Surf A Physicochem Eng Asp 607:125415

Zhang H et al (2021) Encapsulation of curcumin using fucoidan stabilized zein nanoparticles: preparation, characterization, and in vitro release performance. J Mol Liq 329:115586

Zhong J et al (2018) Electrospinning of food-grade nanofibres from whey protein. Int J Biol Macromol 113:764–773

Zhou Y et al (2022) Preparation and stability characterization of soybean protein isolate/sodium alginate complexes-based nanoemulsions using high-pressure homogenization. LWT 154:112607

Ziegler EE (2011) Consumption of cow's milk as a cause of iron deficiency in infants and toddlers. Nutr Rev 69(suppl_1):S37–S42

Zuidam NJ, Heinrich E (2010) Encapsulation of aroma. In: Encapsulation technologies for active food ingredients and food processing. Springer, pp 127–160

Index

Printed in the United States
by Baker & Taylor Publisher Services